Evolution and the Future

BEYOND HUMANISM: TRANS- AND POSTHUMANISM
JENSEITS DES HUMANISMUS: TRANS- UND POSTHUMANISMUS

Edited by / Herausgegeben von Stefan Lorenz Sorgner

Editorial Board:
H. James Birx
Irina Deretic
James J. Hughes
Andy Miah
Domna Pastourmatzi
Evi Sampanikou

Vol. / Bd. 5

Zu Qualitätssicherung und Peer Review der vorliegenden Publikation

Die Qualität der in dieser Reihe erscheinenden Arbeiten wird vor der Publikation durch den Herausgeber der Reihe geprüft.

Notes on the quality assurance and peer review of this publication

Prior to publication, the quality of the work published in this series is reviewed by the editor of the series.

Evolution and the Future

Anthropology, Ethics, Religion

Edited by
Stefan Lorenz Sorgner
and Branka-Rista Jovanovic

In cooperation with Nikola Grimm

Bibliographic Information published by the Deutsche Nationalbibliothek
The Deutsche Nationalbibliothek lists this publication in the Deutsche Nationalbibliografie; detailed bibliographic data is available in the internet at http://dnb.d-nb.de.

Cover Design: © Olaf Gloeckler, Atelier Platen, Friedberg

Cover image: Jaime del Val: Microsexes Metaformance. Photo by Claude Fournier, Toulouse, 2010

Library of Congress Cataloging-in-Publication Data

Evolution and the future : anthropology, ethics, religion / edited by Stefan Lorenz Sorgner, and Branka-Rista Jovanovic, in cooperation with Nikola Grimm.
 pages cm. — (Beyond humanism: trans- and posthumanism, ISSN 2191-0391 ; vol. 5)
 ISBN 978-3-631-62369-5
 1. Human evolution—Philosophy. 2. Human evolution—Forecasting. I. Sorgner, Stefan Lorenz, editor of compilation.
 GN281.E886 2013
 599.93'8—dc23
                                              2013013940

ISSN 2191-0391
ISBN 978-3-631-62369-5
© Peter Lang GmbH
Internationaler Verlag der Wissenschaften
Frankfurt am Main 2013
All rights reserved.
Peter Lang Edition is an Imprint of Peter Lang GmbH.

Peter Lang – Frankfurt am Main · Bern · Bruxelles · New York ·
Oxford · Warszawa · Wien

All parts of this publication are protected by copyright. Any utilisation outside the strict limits of the copyright law, without the permission of the publisher, is forbidden and liable to prosecution. This applies in particular to reproductions, translations, microfilming, and storage and processing in electronic retrieval systems.

www.peterlang.de

Acknowledgements

The editors of the project "Evolution and the Future" thank the publishing house Peter Lang for enabling us to realize this book project. Financial support for the publication came from Kurt Benning, Dieter Mosburger and Siegfried Brenke for which we are extremely grateful, because the publication of this volume would not have been possible without it.

As the various contributions to this volume were originally written for the conference "Evolution and the Future" which took place in Belgrade in October 2009, the sponsors of the conference listed below contributed enormously to the realization of this project. Please accept our best thanks:

Ministry of Science, Technology and Development, Republic of Serbia
Ministry of Religions, Republic of Serbia
Embassy of the Kingdom of Holland, Belgrade
U.S. Embassy, Belgrade
Australian Embassy, Belgrade
British Council, Belgrade
Goethe-Institute, Belgrade
Austrian Cultural-Forum, Belgrade
Yugoslav Cinematheque, Belgrade
NIKOLA TESLA Museum, Belgrade
NGO Responsibility for the Future, Belgrade
Konras Publisher, Belgrade
Haimos Music Society, Belgrade
Serbian Post
Telekom Serbia
Electronic Power Station Serbia
Intesa Bank, Serbia
AIK Bank, Serbia
Konsig Group, Belgrade
Holliday Travel, Belgrade
Hotel Continental, Belgrade

The conference was organized primarily by Branka-Rista Jovanovic in cooperation with the University of Belgrade, the NGO Responsibility for the Future and INES.

The Editors, November 2012

Table of Contents

Acknowledgements 5

Introduction
Evolution Today
Stefan Lorenz Sorgner/Nikola Grimm 9

The Responsible Self
Questions after Darwin
Hille Haker 21

Homo Sapiens, Animal Morabile
A Sketch of a Philosophical Moral Anthropology
Otfried Höffe 35

Enhancement and Evolution
Sarah Chan 49

Ethical Assessment of Human Genetic Enhancement
Nikolaus Knoepffler 67

Evolution, Education, and Genetic Enhancement
Stefan Lorenz Sorgner 85

On the Origins of Modern Science
Copernicus and Darwin
Francisco J. Ayala 101

Technology as a New Theology
From "New Atheism" to Technotheism
Mikhail Epstein 115

Evolution and the Question of God and Morality
The Debate over Richard Dawkins
Dietmar Mieth 129

Evolutionary Theory Applied to Institutions
The Impact of Europeanization on Higher Education Policies
Vojin Rakic 145

Music and Evolution
DNA and the Evolution of Motifs in Beethovens greatest Piano Work
"The Hammerklavier Sonata"
Michael Leslie... 171

Biographies... 187

Introduction
Evolution Today

Stefan Lorenz Sorgner / Nikola Grimm

Darwin's theory of evolution has been one of the most groundbreaking scientific insights during the past centuries. Its importance, relevance and implications have not yet been grasped in all their depth. This essay collection aims to provide some selected contemporary perspectives upon the potential of this theory for our contemporary world and for future generations. The collection does not aim at a comprehensive analysis of the challenges in question, but merely tries to follow and deal with some of the most central traces of the theory.

In the first part of the introduction, a brief analysis will be given of the relationship between the theory of evolution and central contemporary cultural movements, whereby our focus lies on meta-, trans-, and posthumanism, so that the relevance of the theory of evolution for cultural debates is being hinted at. In the second part of the introduction, the various contributions to this collection are being summarized.

Cultural History after Darwin

Darwin's theory of evolution and Nietzsche's perspectivism are two of the most important cultural insights during the previous 150 years. Both have shaped significantly the way many enlightened human beings grasp the world today, and both concepts have brought about a paradigm shift in many Academic disciplines and in the life world in particular in the enlightened parts of this world.

Still, the importance of Darwin's naturalism has long been neglected in many cultural fields apart from the natural sciences, e.g. it did not play a dominant role during postmodernity. The postmodern era can be characterized as the times during which the doubt concerning the truth as correspondence to the world and together with this insight the need for continuous processes of interpretation and play were seen as the basic constituent of our culture. It was the period of time during which a specific aspect of Nietzsche's philosophy, namely his perspectivism, was particularly influential. However, during the previous 40 years, Darwin's theory of evolution and hence, also his naturalism has steadily gained further cultural recognition which can be seen in the fact that Academic disciplines like evolutionary ethics, epistemology or aesthetics have only been developed seriously then. This process is the reason how postmodernity has developed into posthumanism which can be seen as a leading cultural movement in enlightened, mostly Western, countries.

Posthumanism affirms Nietzsche's perspectivism, and hence its postmodern basis, but also regards a naturalist, this-worldy or materialist interpretation of the world as the most plausible one, which is the reason for posthumanists to seriously consider naturalist processes, the theory of evolution and phenomena which used to be interpreted by reference to non-natural entities, like the mind, consciousness, or free will during the humanist era. Posthumanism distinguishes itself from a particular understanding of humanism which can be characterized by the affirmation of categorical dualities, like the one between a material body and an immaterial soul, as it can be found in Christian and Kantian versions of humanism. It is a matter of dispute when this understanding of humanism began to dominate Western cultures. The Stoic times (Sloterdijk) and the Renaissance (Hassan) have been mentioned as potential starting points.

The above mentioned growing cultural importance of naturalist thinking cannot only be seen in the development of new academic fields, but also in public discourses. Most noteworthy here is the fight between creationists and defenders of the theory of evolution concerning the question what ought to be taught in biology classes, which takes place in various parts of the world; e.g. in May 2012, references to evolutionary theory were removed from science text books which are being used in South-Korean schools. Of course, not all defenders of the theory of evolution agree on how exactly the theory ought to be understood. During the second half of the Twentieth century, Lamarckianism which claims that acquired characteristics can be inherited was not too popular. Due to the discovery of epigenetic procedures, it has been taken more seriously again and it has reentered public discourses during the beginning of the twenty-first century. Both the relationship between the theory of evolution and religion as well as the question of epigenetics will be dealt with within this collection.

Not only descriptive aspects of evolution have been discussed in public and in academic settings. Normative questions related to evolutionary theory combined with the progress concerning the options in the field of humanbiotechnologies have developed in an active field of research since the beginning of the twenty-first century. Most of the relevant issues can be summarized under the heading of the enhancement debates within the field of bioethics which is divided into the fields of genetic enhancement (by selection and modification), Cyborg enhancement (e.g. brain-computer interfaces), pharmalogical enhancement (e.g. ritalin, modafinil), morphological enhancement (e.g. plastic surgery) and moral enhancement (e.g.

citalopram and serotonin; oxytocin). This development became possible because a growing number of scholars have accepted the following premises:
- human beings and great apes have common ancestors and are natural beings which have not always been in existence but came into existence as a consequence of natural selection.
- there is nothing in the naturalist world which is not subject to change, and hence it seems highly likely that human beings will die out eventually and/or develop into a different species.
- it seems probable that human beings are neither the highest and best beings on earth, nor the end of the evolutionary chain.

Given that we have also developed the capacity to alter both genotype as well as phenotype of a human being, this raises the question whether the next step in evolution will still depend upon natural selection, and whether this ought to be the case, because human selection might also be a good option for the coming about of the posthuman, e.g. the next step within the process of human evolution.

As a consequence of these developments, it is not much of a surprise that a group of scholars entered both academic as well as popular discourses, who regard it, in some cases, even as a moral duty to promote this development away from the current constitution of human beings. This influential movement calls itself transhumanism.

It is important not to mix up trans- and posthumanism, even though their names sound similar and the concept of the posthuman comes up in various different meanings within both of them. However, these two movements have different cultural roots and uphold different ideals of the good. It is not the case either that both movements do not have anything in common, because both reject Christian and Kantian versions of humanism and hence a dualist understanding of human beings, which implies that human beings are composed of two radically separate substances, namely that of immaterial soul and that of material body. Yet, it is this understanding of human beings which is still dominant in many official documents of Western countries; e.g. in many constitutions, it is being taken for granted that human beings are categorically superior to all other solely natural beings, which they regard as the reason that only human beings are bearers of dignity, and the German constitution represents a paradigm example where only human beings are bearers of dignity and animals are supposed to be treated like things from a legal perspective, even though it is not the case that they are seen as things. It is this traditional Western dualist understanding of human beings according to which only human beings have a categorically special status in the world due to their immaterial rationality which is often part of their rational soul and which is often the reason, too, why only human beings are regarded as beings that are created in God's image. Both trans- as well as posthumanism reject this understaning of humanism. They differ, however, with respect to their main goal. It is the main goal of many posthumanists thinkers to spell out in detail this move away from humanism. On the other hand, most transhumanist thinkers regard the insight just mentioned as starting point for their main goal, namely to reflect upon and to promote the coming about of the posthuman. The following description concerning the cultural embeddedness of various movements away from a dualist concept of humanism is supposed to provide an initial map of some influential contemporary discourses in both the English speaking as well as the continental philosophical realm.

Post-, Trans- and Metahumanism

Posthumanism is a cultural and philosophical movement which is being based in the continental European philosophical and the Anglo-American literary and cultural theory context, and it is intimately connected to postmodernity, because it is an immediate outgrowth of it. In contrast to postmodernity, which is based mainly on perspectivism, posthumanism combines the perspectivism with an affirmation of naturalism, materialism or another type of immanentism. Postmodernists claim that all perspectives are interpretations and apply this insight to various fields of discourses and aspects of the life world. Posthumanists agree with this insight. However, they also affirm that even though immanentism is an interpretation, it is the most plausible one to hold, because it does not depend on metaphysical, two-worldy entities with which we cannot be acquainted immediately.

The term "posthumanism" was coined initially by Ihab Hassan in the article "Prometheus as Performer: Toward a Posthumanist Culture?" from 1977. The leading proponents of posthumanism so far are mostly philosophically minded literary critics and cultural theorists, like Donna Harway who wrote "A Cyborg Manifesto: Science, Technology, and Socialist-Feminism in the Late Twentieth Century" (1985, updated version 1991) and Katherine Hayles who wrote "How We Became Posthuman: Virtual Bodies in Cybernetics, Literature and Informatics" (1999). A literary minded philosopher like Peter Sloterdijk can also be seen as a posthumanist thinker, and he used the term posthuman in some of his writings, e.g. in his infamous essay "Rules for the Human Zoo" from 1999. Also scientifically minded philosophers such as Francisco Varela, Evan Thompson and Humberto Maturana can be seen as associates of the posthumanist project. Their book "The Embodied Mind: Cognitive Science and Human Experience" from 1991, or Varela's and Maturana's "Autopoiesis and Cognition: The Realization of the Living" from 1980 represent many traces of what posthumanism stands for, e.g.

Introduction – Evolution Today   13

plurality, perspectivism and immanentism. It is their way of dealing with the theory of evolution which reveals the relevance of this topic for the posthumanism. The geneticist Eva Jablonka can be seen as being related to posthumanism from a scientific perspective, too. Her book "Evolution in Four Dimensions" coauthored together with Marion Lamb and published in 2005 stands for posthumanism within evolutionary biology. Given the wide spectrum of researchers and thinkers who are related to posthumanism, it can be described as a diverse and a contemporarily strong cultural movement.

The next such movement is more unified, but also intimately related to the question of evolution, because the thinker who coined the term transhumanism belongs to the English language tradition of the theories of evolution: Julian Huxley. The Darwin supporter Thomas Henry Huxley was his paternal grandfather, and the author of the novel "Brave New World", Aldous Huxley, was his brother. The lesser known half brother Andrew Huxley, a biologist, won the noble prize. Julian Huxley was a member of the British Eugenics Society and even their president for several years. In 1957, he coined the term "transhumanism" in his monograph "New Bottles for New Wine" by stressing the need of human beings to transcend themselves by means of the usage of science and technology. However, the contemporary concept of transhumanism is more closely related to the ideas of the Iranian futurist Fereidoun M. Esfandiary, better known as FM-2030, who wrote the "Upwingers Manifesto" (1973) and the book "Are You a Transhuman?: Monitoring and Stimulating Your Personal Rate of Growth in a Rapidly Changing World" (1989), and his former partner Natasha Vita-More who wrote the "Transhumanist Arts Statement" which came out in 2003 and which is a revised version of the "Transhuman Manifesto" from 1983. Nowadays, Natasha Vita-More is married to Max More, whose essay "Transhumanism: Toward a Futurist Philosophy" from 1990 was particularly influential in forming the currently dominant understanding of the term. However, his libertarian sympathies have also been criticized by many fellow transhumanists who are associated more closely with a social-democratic understanding of transhumanism which is being represented best by James Hughes within his monograph "Citizen Cyborg: Why Democratic Societies Must Respond to the Redesigned Human of the Future" which can also serve as an excellent introduction to transhumanism. A further significant step for the cultural influence of transhumanism was the foundation of the "World Transhumanist Association" in 1998 and their "Journal of Transhumanism" which was renamed "Journal of Evolution and Technology" in 2004 and transferred to the "Institute for Ethics and Emerging Technologies" (founded in 2004 by Nick Bostrom and James Hughes).

In contrast to posthumanism, transhumanism is closely related to the English speaking world of the natural sciences, analytic (bio)ethics and utilitarianism, and is characterized strongly by the affirmation of the use of sciences and technologies in order to promote human capacities. The human capacities mentioned most often in this context are intelligence, health, memory, the capacity to concentrate, and the prolongation of the human health span which is different from the human life span because it stresses the relevance of the period of time in which one lives healthily. However, a great variety of capacities are being mentioned among transhumanists when these issues are being discussed. What is most relevant from the transhumanist perspective is that the development of the capacities leads from human beings via the transhuman to the coming about of the posthuman. However, it is a matter of dispute among transhumanists whether the posthuman is still a member of the human species, but has one capacity which goes beyond the capacities of currently living human beings (Bostrom) or whether the posthuman is actually a member of a new species (FM-2030). As the enhancement of human capacities is one of the key features of transhumanists, their positions represent strong voices within the bioethical enhancement debates which have taken place since the beginning of the 21st century. Due to the relevance of these debates concerning the possibility to enhance evolution, this movement is relevant for the debate concerning evolution and the future, too.

Most recently, the Spanish artist Jaime del Val and the German philosopher Stefan Lorenz Sorgner recognized the need to bridge the gap between posthumanist and transhumanist discourses and developed a philosophical and artistic attitude which moves *beyond* a traditional dualist version of humanism, but which also lies *in between* trans- and posthumanism. They named their own approach metahumanism, because "meta" means both "beyond" as well as "in between" and hence covers both aspects of their initial intention whereby Sorgner is more closely related to the English language tradition and del Val more closely to the French philosophical world. However, due to their common high estimation of Nietzsche and his perspectivism, their disrespect of paternalistic structures and their high evaluation of radical plurality, they managed to form some guiding principles which both of their works have in common.

An important step for realizing an exchange or bridge between post- and transhumanism was a special issue of the "Journal of Evolution and Technology" which was dedicated to the topic "Nietzsche and European Posthumanisms" (2010) in which articles critically dealt with an article by Sorgner concerning the relationship between Nietzsche and transhumanism from 2009. This exchange continued in the Fall 2011 issue of the journal "The Agonist" published by the Nietzsche Circle/New York in which several leading Nietzsche scholars responded to the ongoing debate. In Sorgner's further response published in this special issue, he put forward a bioethical position which can be seen as a weak version of transhumanism which is deeply embedded in the continental philosophical tradition

Introduction – Evolution Today

and thereby also stands for a weak version of posthumanism. A central issue of the debate was the question concerning the relationship between Nietzsche's overhuman, the transhumanist concept of the posthuman and the ethical question concerning the moral obligation to promote human capacities. Again, the question concerning the future of human evolution was a focal point of the philosophical debate.

Evolution and the Future

The importance of Darwin's theory of evolution for contemporary cultures, and philosophical and artistic reflections in the 20th and 21st century and with respect to some strong cultural movements of today reveal the central motivation of this book project. The project unites selected papers of the conference "Evolution and the Future" which took place in October 2009 in Belgrade and which was organized by Branka-Rista Jovanovic in cooperation with the University of Belgrade, the NGO Responsibility for the Future and INES. Leading scholars from a diversity of cultural, ethical, and scientific backgrounds consider the relevance of the theory of evolution and the potential impacts of this theory and its implications for future developments. Within this collection the fields "Anthropology, Ethics, Politics, Religion and Science" were considered in particular detail.

Anthropology

At the beginning of this collection and as an initial input into the wide discussion, three perspectives are being presented concerning the basic entities which have to be dealt with when considering evolution: ourselves, the anthropos, human beings. In the article "The Responsible Self – Questions after Darwin" the Catholic theologian Hille Haker presents lines of thought concerning a new concept of the self. In the first part she summarizes existing concepts of the self, whereby she identifies three basic characteristics of the notion of the self and identity in the 20th century: subjectivity, practical identity and the moral self. In part two she considers the relationship of the characteristics by focusing on the question of self-identity, whereby she is pointing out two main concepts, which are important for the constitution of the self: "belonging as recognition" and "selective narration". Based upon some insights of the neuroscientist Jean-Pierre Changeux and the philosopher Paul Ricoeur, Haker discusses the place of one-self finding itself in the dialectic of social constitution and the self-discourse in narration. In between these two characteristics, Haker emphasizes the development of a moral identity in the third part of her paper by stressing the accountability for ones actions despite the dependence on others in a certain context. In the fourth part the author puts forward a concept of the "responsible Self" by considering further the interaction of biology and morality. According to Hille Haker, the theory of evolution has to accept that natural selection is based on the "hermeneutic principle of narrative selection", which is constitutive for a whole human being, being a moral agent in the end.

The theory of evolution provides the framework of Ottfried Höffe's argument of his article "Homo Sapiens – Animal Morabile". There, he focuses on the question why and how natural evolution creates a moral being like the human, which he develops in seven steps within the field of "philosophical-moral anthropology". Departing from the existing skepticism concerning such an anthropology, he argues for its possibility and shows its internal logic. Thereby, he considers the biological basis of morality and also ethical naturalism. As a fifth step Höffe explains that humans have a non-specific and potentially dangerous overdrive that needs to be directed, which takes him to the question whether animals can be moral. In a provisional conclusion he explains two ways of grasping morality: Firstly, he regards morality as possible due to the "intelligence and openness to the world" of human beings. Secondly, he sees morality as necessary because of the human "overdrive and openness to the world". Hence, morality is not given by birth but it is linked to the development of one's own personal standards and powers during a lifetime.

Ethics

Both animals and human beings are being considered in Sarah Chan's argument of her article "Enhancement and Evolution". In three parts she puts forward reasons in favour of the enhancement of humans as well as of animals. By analyzing general main issues of the enhancement debates in part one, Chan stresses the need for human enhancement, including even "a moral imperative to use it for the benefit of future generations". The future of evolution with respect to human enhancement is the main topic of part two. Instead of these processes leading to a loss of humanity, Chan sees the chance to redefine what this word will mean by developing a concept of human being, who is able to reflect about himself and shape and change the world according to his wishes. Hence, she does not regard the effects of enhancement on evolution as dangerous, but defends them as process, which lies in continuity with human nature. In the context of enhancing evolution, the third part is dedicated to the question concerning the

Introduction – Evolution Today 17

enhancement of animals. Hereby, she finds positive moral reasons to promote such processes, as she points towards the obligation not to limit enhancement to one biological species only, but to include creatures who share "the qualities we value". Given this approach, it might be even our duty to make animals capable of deciding for their own best.

Nikolauf Knoepffler's contribution, on the other hand, is focused solely on the human species with respect to the debates concerning genetic enhancement. The title of his article represents its goal: to give an "Ethical Assessment of Human Genetic Enhancement". After providing the reader with a definition of the "meaning and limits of the topic" to explain the framework of his argument, Knoepffler focuses on the moral concept of human dignity, which is the central norm of bioethical discussions in Germany and in many other international charters and laws. Then, he shows how to make normative judgments concerning ethical questions by dealing with conceptual categorizations of various types of genetic enhancement. In the end, he develops a sketch of how to proceed concerning genetic enhancement on the basis of human dignity and applies his method at three cases: Improving the human eye, improving memory performance and changing the behavior of a pedophile. By relating each of the examples to his moral basis, he reveals a way of considering the case in question and hints at a possible solution.

In contrast to the dominant evaluation of genetic enhancement in Germany and in many other countries, Stefan Lorenz Sorgner's main goal is to provide some reasons for holding that "genetic enhancement does not have to be seen as morally problematic" in his article "Evolution, Education and Genetic Enhancement". Thereby, he employs the method of revealing structural analogies between traditional procedures and genetic enhancement, and he regards genetic enhancement as the most promising technology for having evolutionary consequences. His argument progresses in three steps. Firstly, he describes various concepts of enhancement, whereby he explains why he focuses on heteronomous types of genetic enhancement during the following parts of his paper, in particular genetic enhancement by selection and by modification. Secondly, he explains why genetic enhancement by means of selection and choosing a partner for procreative purposes ought to be seen as structurally analogous, and thirdly, he puts forward some reasons for claiming that the same applies to genetic enhancement by means of modification and classical education. According to Sorgner, it is highly likely that genetic enhancement does not have to be morally problematic, as classical education and choosing a partner for procreative purposes do not have to be morally problematic procedures.

Politics, Religion, and Science

The third section begins with the article "On the Origins of Modern Science: Copernicus and Darwin" by Francisco J. Ayala in which he shows parallels between the Copernican and the Darwinian revolution, as he aims at a reinterpretation of the meaning of these two events. Darwin follows and finishes a process which has been initialized by Copernicus by extending the reduction of natural laws to human beings. In contrast to many scholars who regard the theory of evolution mainly as an insult to human self-understanding, like Freud, whose psychoanalysis is considered as the third significant insult, Ayala appreciates both events as shades of one process towards the scientific worldview which we share nowadays. Then, Ayala concentrates on Darwin's theory and his main work, the "Origin of Species", to explain its essential line of thoughts and to help grasp evolution as a non-directive, but nevertheless creative process without the need of a supernatural designer.

Michael Epstein, on the other hand, does not exclude the possibility of a supernatural designer, and relates the question of evolution to the one concerning religion and belief in general. His article "Technology as a New Theology – From "New Atheism" to Technotheism" is split up in five parts. Firstly, Epstein criticizes the genocentrism Dawkins has developed in his book "The God Delusion". He regards Dawkin's view as a "new religion", in which God's place is being replaced with that of genes. In contrast to existing religions, this one is supposed to be more reductive and simplistic, as it reduces human beings to "slaves of genes". Epstein furthermore stresses that the loss of morality is an immediate consequence of the theory of the selfish gene. In the second part he argues in favor of the existence of God as the creator of the Universe. In contrast to the understanding of technology as process which leads people away from believing in God, Epstein regards the development of technologies and sciences as an easier way of approaching God: Being a human creator facilitates grasping the concept of being created oneself; to experience the possibilities of technologies enables people to open up for the idea of God's omnipotency. To extrapolate from science to God is a new way of grasping God. By analyzing some of Bostrom's reflections in part three, he develops this line of thought further. In parts four and five, he stresses the probability of there being a God, and explains why this insight is consistent with our concept of technology. To integrate this understanding of technology within a religion would complete and enrich both fields, from Epstein's perspective, and it is this wish with which he ends his paper.

Dawkins theories are also the main focus of Dietmar Mieth's article "Evolution and the Question of God and Morality – The Debate over Richard Dawkins". His

Introduction – Evolution Today

argument is divided up into two steps. Within the first part of his article, Mieth introduces new lines of thoughts into the debate concerning Dawkins' position. In the second part, he stresses the need for further critical reflections on science. In his criticism of Dawkins point of view, he reveals deficits within his arguments, although he admits the relevance of some of Dawkin's critical arguments against religion, e.g. the reversal of the authoritarian fallacy. Mieth stresses the need to be fair by pointing towards the "religious heritage of humanist thought". The second part of his text refers to the role of prejudices within biosciences to demonstrate the science's "belief" it has in itself. Thereby, science seems to be self-contradictory: it is moved by the belief in an uncertain knowledge but condemns any other types of belief. Mieth stresses the relativity of knowledge and argues for freedom and responsibility in scientific research. He concludes by referring to incoherences which are typical for a modern scientific approach and which manifest themselves in Dawkins' reflections, too.

The phenomenon of evolution within the political sphere is being presented by Vojin Rakic in his article "The Evolutionary Theory Applied to Institutions". He particularly concentrates upon the impact of European integration on higher education policies, whereby his specific focus lies on six exemplary states of the European Union: Germany, the Netherlands and Belgium/Flanders, Great Britain, Sweden and Finland. Firstly, he exemplifies different concepts of convergence and divergence, brief change, to be able to conclude that the states rather converge than diverge. From this description he develops a "general scheme" of possible reasons. The next step within his paper is the question, whether significant changes have occurred in the six states and to what extent this was the result of EU policies. In this context, Rakic provides us with a general overview of the development in each of the states. In the following chapter, Rakic checks the plausibility of his "scheme" with respect to two existing theories that explain imitation mechanisms. In a final epilogue, Rakic refers to evolutionary theory in particular. By means of the analogy between education policies and life in the context of evolution, he reaches the following conclusion: The adaption of higher education policies to a competitive and globalized environment does not exclude the influence of political leaders of a system; life does not depend on evolutionary selection but allows the existence of a creator. This insight stresses the compatibility of evolutionary theory and intelligent design which is the main conclusion of this article.

A Musical Epilogue

The final section of the collection presents a musical epilogue entitled "Music and Evolution – DNA and the Evolution of Motifs in Beethoven's greatest Piano Work 'The Hammerklavier Sonata'" written by the Australian-German concert pianist Michael Leslie. Leslie presents Beethoven as an architect of music, who manages to put together a complex and nuanced work of music with the Hammerklavier Sonata, which in certain respects has never been reached by anyone else again, not even Beethoven himself. By means of direct references to parts of the score, Leslie analyzes this composition which reveals the subtlety of Beethoven's precisely organized work. With his article, he underlines the high level of organization of this work by examining the seemingly random episodes in the introduction, which prefigure the "theme of the final movement, the subject of the fugue". Episode after episode the pianist discovers non obvious connections which actually form a complete and highly complex musical work. Even unprepared listeners become immediately affected by this extraordinary piece of music, according to Leslie, although its deeper glory is being revealed only to experts, the initiates, who know the relationship of the initial episodes to the sections of the fugue. This knowledge allows initiates a more profound aesthetic perception in comparison to that of unprepared listeners.

Conclusion

The four sections of this collection, listed above, reveal some of the central challenges related to the question of evolution: 1. What are the fundamental processes of human evolution?; 2. How is it to be evaluated morally that human beings shape their own future or bring about trans- and posthumans by means of enhancement technologies?; 3. What is the relationship between the theory of evolution and belief in God?; 4. What is the relevance of the arts from an evolutionary point of view? These questions provide a general framework for future academic and public discourses concerning the theory of evolution. Hopefully, this collection manages to provide further stimulus for the various present and future debates on all of these timely, highly important, and fascinating topics.

The Responsible Self
Questions after Darwin

Hille Haker

Introduction

The concept of identity is one of the most complex concepts in philosophy. It not only explains the constitutive relation between two entities, either as being identical or different, in logic and semantics, and it not only tries to give order to an otherwise contingent reality of the world, as reflected upon in the tradition of metaphysics – it also addresses the question of personal identity, and/or the identity of the self.

The Philosophical Discourse: Subjectivity, Existentiality, and Morality

Although I do not intend to analyse the philosophical concepts of the self either in the ontological, the empirical, the idealistic, or the phenomenological traditions, it is necessary to summarize the relation between these traditions and contemporary thinking in a few words.

From the 16th century to the end of the 19th century, the notion of the self was the safeguard against utter scepticism or even relativism in Western philosophy – either as image of God in theological reasoning, thus participating in eternal and divine knowledge, or as basis for the possibility to make claims that are ontologically true, or as the constitutive subject of consciousness defining the epistemological status of knowledge. Even though the paradigm shift from ontology to epistemology had long been prepared, it was not until the late 19th and then the 20th century that the self was radically questioned from psychology, biology, sociology, and finally philosophy itself. In Western thinking, which is the referential context of this lecture, the status of the concept of identity is entangled with the status of religion: As Charles Taylor has shown convincingly, the loss of religion as the decisive source of the self has far-reaching effects – as much for the self as for religion. Here, let me just state the results with respect to the concepts of the self.

Three basic characteristics accompany the notion of the self and identity in 20th century philosophical thinking up to the present. These are not radically new in themselves but, taken together, they bring about a shift in the conceptualization of personal identity especially with view to the ethical reflection:

First, *subjectivity*: while substance philosophy claimed that an entity exits in and of itself – and can be examined as such, the turn to subjectivity is – among other things – a breach with the concept of a possible cognitive representation of these entities. In the first half of the 20th century, the concept of subjectivity is connected to the insight of phenomenology, namely the claim that any knowledge is dependent on the way it is seen by a self that 'intends' the world in all her operations. Kant's program is transformed by Husserl and Heidegger, because it would not make exactly this last radical step to constructivism of any act of consciousness. The self of the phenomenological tradition, we may say, is at the same time radical subjectivity while maintaining the claim of an outside world, via the construction of *meaningfulness* of being through the operations of intentionality.

While this is complex already as an epistemological concept, it is complicated further when applied to the self. Phenomenology claims that the self is partly the object of the world, just as other objects; it can be analysed in similar intentional acts of appropriation and analysis as all the other objects of reasoning, which in the 20th century become dominantly the object of empirical science, with many new and exciting insights gained from neurosciences. However, different from all these other objects in space and time, the self is an embodied self that cannot relate to itself in the same way as to other objects of the world. Hence, a fundamental dualism of perspectives is revealed: on the one hand, we address ourselves as bodies, and on the other hand, we address our-own-body as selves, i.e. as we experience ourselves as embodied selves. These two perspectives or discourses cannot be translated into the traditional language of body and soul, or the material and immaterial, because in the second perspective, the corporeal dimension is maintained, however different to the "body-as-object" perspective.

Because of the constitutive embodiment, the gulf between *being* oneself and *seeing* oneself with the eyes of others cannot be bridged. Furthermore, because of this, the self cannot be reduced to the 'self as body', or even to the representations of physical, psychic or mental states expressed in brain activities. They do not touch upon the dimension of being oneself as an embodied self. It is a gap that separates empirical sciences of today, too, from the humanities (though not the social sciences).

The turn to subjectivity does not mean that we are all monads without windows. Even though it is true that I cannot experience your pain or your passion, there are bridges between the self and the other, and even more so, as far as the self is to become conscious of oneself, he or she is to make use of operations stemming from the sphere of the "other", most importantly, stemming from a language he or she cannot fully invent but rather discovers.

If the self is to learn about her- or himself, it needs to do this through different kinds of mediations or operations, namely with the other and the world. By way of these symbolic mediations, the self is dialectically constituted as much as it constructs her own identity. We will see that – if this process of symbolic mediation is taken serious – phenomenology must be transformed into a specific hermeneutic of the self.

The phenomenological turn changes the concept of religious identity, too. Subjectivity as presupposition of the turn to the phenomena renders a simple relation between man and the divine impossible. Far from the traditional search of evidence of God in the world that can be discovered by metaphysical reasoning based upon the human participation in the divine knowledge, and also going beyond the dialectic of the transcendental self and transcendence, God is in the world only through the self's creative imagination, as Richard Kearney puts it.

Most approaches in philosophy in the 20th century, however, are laid out more or less independent of this theological thinking, because the constitutive source or point of reference for the necessary mediation of the self is no longer God; furthermore, the epistemological paradigm has changed through the modern development of the sciences. Particularly evolution biology and, later, life sciences, genetics, and neurosciences push the understanding of embodiment – to be a body and not only to have a body – more and more to background, in favour of the reductionist construction of the concept of 'life' as biological matter.

Second, *practical identity*: While the theoretical reflections on personal identity is continued in many different twists and turns, the second notion of the concept of the self, connected to subjectivity but focussing much more on the concept of autonomy as self-determination, concerns the practical identity just as much as the theoretical concept of the self. While the Lockean tradition focuses on the identity of the self as a continuous self throughout time, and on a practical self that can be held accountable for her actions, 20th century's continental philosophy takes yet another turn, stemming more from the Hegelian tradition: Beginning with Hegel, inter-subjectivity and social interaction become a central feature of self-development. Together with Kierkegaard, it is Jean-Paul Sartre who fully develops a concept that can be identified as the *ethical existentiality* of the modern self. Based on both the social constitution as well as on radical subjectivity of the self, freedom is a chance as much as it is a curse for the self: the necessity to make choices, here and now, defines us in our freedom, but on the other hand, it is only through these choices that we can develop an 'authentic self' as being or becoming oneself: proof of our individual identity, an identity that is not fully transparent but that is nevertheless engaged in one's own experiential search for the good life, based on freedom.

Third, the *moral self*: The third notion also concerns the practical concept of identity, but it turns to *morality* in the normative sense of accountability for one's actions in a much greater range that this is the case before. While it is certainly not remarkable that there is a tight relation between the concept of the self and morality, it is worthwhile to take a closer look how exactly morality becomes a defining characteristic of personal identity in the 20th century. For this, however, we need to take on board the identity discourse in psychology and sociology that in the last century developed together and in connection with the philosophy of the self on the one hand, and social philosophy on the other.

Similar to practical philosophy, psychology and sociology are focused on the practical self. But they are much more interested in the development of personal identity through interaction with others and, in fact, in all the different social spheres with which persons are involved. Again, it is not possible to elaborate on all the different approaches, from Sigmund Freud to Erik Eriksson or Jean Laplanche, from G.H. Mead or Anthony Giddens, and from Jean Piaget to Lawrence Kohlgan and Carol Gilligan, to name but a few. I would rather like to summarize a few characteristics here, again, under the headline of two main concepts, namely the concept of belonging as recognition and narration as the necessary mediation form to give meaning to oneself.

Self-identity and the Development of Moral Identity

In 2000, an experimental and experiential dialogue was published. It had been held by neuroscientist Jean-Pierre Changeux and Paul Ricœur (Changeux/Ricœur 2000). In this protocol of a dense and difficult conversation, the clash, but also the attempt to overcome the divisions between empirical science and philosophy are revealed. While the neuroscientist claims that mental states, including the subjective experiences of one-self will be "naturalized" so that they are accessible by other persons, Ricœur insists on the constructive nature of the experiments. He claims instead that the fundamental dualism of perspectives cannot be overcome and synthesized in a third perspective that takes advantage of representations enabled by the neurosciences. Let us listen for a moment to their conversation, newly arranged (and abbreviated) by me, according to my interest here to make the point of their departure:

The Responsible Self

Changeux: "What I propose to attempt is a naturalization of intentions that takes into account both the internal physical states of our brain and its opening to the world with reciprocal exchanges of meanings, exchanges of representations oriented as much toward perception as to action. Today, I think that observational methods make it possible to obtain physical facts about subjective psychological states."(Changeux/Ricœur 2000, 67).

To this proposal that is very optimistic about the prospects of such a "third perspective" after dualism, Ricœur responds the following:

Ricœur: "I do not at all exclude the possibility of progress in scientific knowledge of the brain, but I wonder about our understanding of the relationship between such knowledge and actual experience. ...
    My question really has to do with whether one can model subjective experience in the same way one can model experience in the experimental sense. Can the comprehension that I have of my place in the world, of myself, of my body and of other bodies be modelled without doing damage to it – epistemological damage that entails a loss of meaning?" (Changeux/Ricœur 2000, 69).

Shortly before, Ricœur indicates his alternative view that paves the way, too, for my own investigation here:

"Ordinary experience does not exactly coincide with what scientists include under the term introspection. Language forces us to escape private subjectivity. It is an exchange that rests on several assumptions: first, the certainty that others think as I do, see and hear as I do, act and suffer as I do; next the certainty that these subjective experiences are at once unsubstitutable (that is, you cannot put yourself in my place) and communicable ("Please – try to understand me!"). One may speak intelligibly of having comparable impressions while watching a sunset, for example. There is indeed such a thing as mutual, even shared, comprehension. This sort of comprehension is, of course, open to doubt. Misunderstanding is not only possible, it is the daily bread of conversation, but it is precisely the function of conversation to correct misunderstanding as far as possible, and to seek *Einverständnis* of which Gadamer and the partisans of hermeneutics speak. There is a hermeneutics of daily life that gives introspection the dimension of interpersonal practice." (Changeux/Ricœur 2000, 68)

Let me explore a little what Ricœur means when he indicates the inter-subjective, interpersonal dimension of introspection, or rather knowledge of oneself as embodied self in a world he or she shares with others.

Personal identity is shaped in a complex process that psychologists and sociologists call personal development or socialisation. In view of this research, the self develops from and in a web of personal, family, and social relations that he or she does not choose. Furthermore, a child is completely dependent on being cared for by others, not only in physical terms but also psychologically (Laplanche). By way of these relations, persons are defined by categories to which they belong, and at the same time they develop a sense of belonging that is an important starting-point for the choices they make: the objective side and the subjective side may not coincide; in the development of personal identity, however, they are interrelated dialectically, and together they define identity as the social identity of a person. For this, the neurosciences can help us find the traces, the "imprints" of this painful integration process, both in the study of brain activities and of the specific individual memory-structure of a person.

Identity is developed by way of and in conflict with the identities others ascribe to us. This begins with the simple fact that persons make use of language that they discover rather than creating it. In self-reflection, for example, we use the name that other people have given us. A name is not chosen, but rather appropriated and at the most given meaningfulness in the development of identity. It is often the case that with the help of a person's name continuity between generations is created which is in turn intended as a source of identity. In the tradition of G. H. Mead, however, the dialogical constitution of identity implies even more. Mead created the concepts "I" and "me" in order to express two different dimensions of personal identity. The "I" perspective indicates the individual self-perspective which in the phenomenological tradition is further distinguished into embodiment and reflective dimensions. The "me" perspective on the other hand is the perspective from outside, the way I experience myself in the view of others. What we are and who we are is often the result of an overwhelming force of external ascriptions and only to a small extent the success of our individual search for identity. Beyond this, identity is a dynamic concept, so that it is possible to speak of learning experiences and self-corrections.

Hence, although it seems to contradict the turn to subjectivity within the theoretical identity discourse, the social self is much more the object of socialisation than being the subject of it. Identity is ascribed to us by multiple categories: sex, class, race, or religions are but a few of these categories. Whereas some of these categories or rather, spheres of belonging, are not chosen, others are. The dialectic between the subjective appropriation and integration and the objective ascription, however, cannot be escaped. The development of self-identity has been described in different terms. Psychological and sociological theories have emphasized that the work of identity (Keupp) entails not only role-taking but also the handling of different roles in the different spheres. Integration of these social activities does not end after the stage of adolescence, as former theories assumed; rather, the 'patchwork identity' needs to constantly and flexibly successfully integrate different performances of the self. What seems to be quite clear loses its 'natural' basis if we look a bit closer. Let me give some examples:

> Sex: A child is born into a particular family: whether it is a girl or a boy is determined by biological, genetic, and social standards as the history of intersexed persons may well tell us. Even though it seems to be the case that one's sex belongs to the non-chosen characteristics, this is only the case as long as an individual can be classified according to the standard rules of masculinity and femininity. The social – normative – rules that are as much rules or standards of normalisation, as Foucault says, appear only in the reflective view of the self or so-

ciety: this is all the more the case when these standard cannot be met, of when they do not mirror one's own experiences. Jeffrey Iogenides has told a heartbreaking story of the search for identity in his novel "Middlesex", interestingly linking the story of coming to a country as stranger with the search for an individual identity beyond the standard normalisation processes. Even though in this novel "Middlesex" is but the name of a city, it becomes the metaphor for the struggle of identity. Gender studies of the last decades have made clear that although there is a biological basis of one's sex, this is surrounded by culturally-dependent norms that define how this biological basis is to be interpreted (Faust Sterling).

Kinship: While kinship seemed to be a non-chosen, biological fact, cultural history and ethnology have revealed the arbitrariness of kinship, too. With the development of assisted reproductive technologies, this goes far beyond the traditional boundaries: it is not clear anymore whether we speak of the genetic, biological, or social parent of a child; hence, a child might ask where the sources of her belonging really are and, at times, even *choose* among these different parents (O'Neill).

Ethnicity: In his bestselling novel 'The Human Stain', Philipp Roth tells the story about a 'white' person who in fact is 'black' according to his heredity. While he attempts rather successfully to hide this 'stain' in the social world, he still is unable to forget his past, and the betrayal of his origin. Racism discourse has shown convincingly the social construction and normative force of these categories.

Nationality: The arbitrariness of classification is also obvious in the case of national identity where some states choose the *ius sanguinis*, and others the *ius soli* as classification of membership, which is yet another category of belonging.

Religion: Last not least, religion is a category that seems to belong to the voluntarily chosen affiliations, and yet, most people grow into a religion just as they grow into a particular clan, culture, or class. One result of the development of a self-identity is the reflective attitude towards most of the categories, even though some are more easily changed or denied than others. At present, for example, we observe the struggle for a redefinition of religious identity in many countries, cultures, but also in different religions, or different currents within given religions. This struggle concerns in part the question of religious freedom, for example in the case when conversions are not granted legally and/or socially.

Given that a person needs to integrate the multiple and plural identity categories, together with their intersections, it should be marked that it makes a difference whether we speak of the identity of a self being, for example, described as a religious person, or religious identity. As Amartya Sen puts it, it is a mistake to see human beings in terms of only one affiliation:

> "... to take that [classification] to be the overarching basis of social, political, and cultural analysis in general would amount to overlooking all the other associations and loyalties any individual may have, and which could be significant in the person's behavior, identity, and self-understanding. The crucial need to take note of the plural identities of people and their choice of priorities survives the replacement of civilizational classifications with a directly religious categorization." (Sen 2006, 60f).

Addressing the "Muslim identity", rather than taking the identity of person who happens to be "a Muslim", Sen holds, would ignore this plurality of affiliations and belongings:

"To focus just on the simple religious classification is to miss the numerous – and varying – concerns that people who happen to be Muslim by religion tend to have." (ibid.)

Arguing for a liberal concept of identity, Sen urges us not only to respect the plurality of one's identity but also to consider the choices each one must make, either to commit to or to remain a member of a certain religious affiliation. Moreover, and just as important, it is not determined by others but the self's own decision how to act within the range and plurality of a given religion, for example, to which of the currents within one's religion one wants to belong, and when one needs to dissent from policies taken.

Finally, successful self-identity is dependent on the recognition by others that defines the status of belonging. By way of this necessary recognition – which mediates, maintains and transforms social norms, the self develops his or her self, i.e. a self that is able to make deliberate choices about her own life and identity.

Today, identity concepts are certainly based upon this dialectic of social constitution, self-identity and the struggle for recognition. Certain "imprints" (to use the language of analogy to the sciences) will be traceable in brain imaging and other technologies. Interestingly enough, it may be possible to find a "third way" beyond naturalisation and dualism, when we consider this dialectic as one that today must include the insights of neurosciences.

I will now turn to the second characteristic of the self-discourse in the humanities, namely narrativity.

With the expression "entangled in stories", Wilhelm Schapp summarises in his book that carries the same name what he views not only as the a priori structure of every act of perception and understanding, but also as the prerequisite for feelings and acts of the will (Schapp 2004). By means of this theory, Schapp radically refers the consideration of personal identity – and not only this, but also the consideration of the human relationship to oneself and to the world in general – to the narrative structure on which the identity is based. Schapp's theory therefore makes a very wide-reaching epistemological claim: neither a theory of knowledge, nor of behaviour can be properly constructed independently from the human being's entanglement in stories. Understanding oneself for Schapp means understanding one's story and one's "self-entanglement" in it; to understand other people requires understanding them in their own stories of self-entanglement, understanding these as "other's stories", which is only possible by entering into a relationship with the other's story, with the other person, and beginning a new story.

Hence, any new story is not only bound up with other stories but also with the story teller, to whom it indirectly points, and with the listener, who grasps its content in a process of active understanding. The listener or reader refers to the story being told and in understanding continues it by comparing it with other stories and constructing a context between story teller, story and herself. Entanglement in other stories is always related to self-entanglement, just as Kant's transcendental apperception, his claim that a *Vorstellung* or presentation is accompanied by the presentation of "I think."[1] The self's reflection of him/herself, however, that is Schapp's claim, must be conceived not just as a presentation but a narration:

In conclusion, although physical self-experience is immediate and individuality stands for uniqueness and incalculability, the identity of a self is constituted by encounters with the other and the world, and these encounters can only be comprehended as experiences when they are articulated and interpreted – hence the turn to narration. The choices someone makes are not to be seen as 'just' based upon natural instincts *or just* individual autonomy but rather as result of a process of self-reflection that is tightly connected to narration; the identity of a person grows out of the story she tells, revises and varies under the impression of new experiences – a process I am tempted to call *narrative selection.*

The stories of oneself are embedded in the knowledge systems, social relations and normative orders of, among others, particular cultures, legal traditions or religions, from which we all take up words, symbols, rituals, and actions, in order to give our identities new constructive shapes. Together, they create spheres of belonging, or even collective identities that cannot be analyzed without examining the 'other side' of identity, and certainly identity politics of groups, which means that the formation process of identity presupposes not only the distancing from 'others' but also at times their exclusions.

Moral Identity

The last two decades have made this turn to the narrative self in an unforeseen way. While the liberal tradition sees the telos of successful identity in the reciprocal recognition of free agents who have developed a reflective identity that enables them to make choices based on their individual autonomy, communita-

---

1 „Das Ich denke, muss alle meine Vorstellungen begleiten können; denn sonst würde etwas in mir vorgestellt werden, was gar nicht gedacht werden könnte., welches ebensoviel heißt, als die Vorstellung würde entweder unmöglich, oder wenigstens für mich nicht sein." (Kant, Kritik der reinen Vernunft, B 131f).

rian philosophers, quite a few of them Christian and Jewish, such as Alasdair MacIntyre, Charles Taylor, and Michael Walzer, have emphasized the necessity of structures and spheres of *belonging* as spheres of recognition.[2] Social movements such as the women's movement, the civil rights movement, or the movement of indigenous people have all stressed the importance of recognition. The self seeks and grants recognition in the different spheres Axel Honneth, among others, has addressed: above all the family, civil society, and the political-legal domain of the public.

While more traditional theories of identity could speak of standard biographies, a narrative unity of biographical integration, or processes of adaptation to a given social sphere, newer theories rather speak of post-teleological, plural and open life-forms that cannot – and in fact should not – be fixed in particular patterns of narratives. This then is much more compatible with Darwin's view of the non-teleological evolution theory that nevertheless can trace evolution, i.e. changes and transformations. One of these transformations concerns the dimensions of morality.

Searching for meaningfulness of one's life may easily be seen to be a desperate desire, or even a burden. Metaphors such as 'thread of life' suggest that in deed we have one big story to tell about ourselves. What does it mean that this assumption has been questioned throughout the 20th century? What effect does this have on Schapp's theory of entanglement and hermeneutic phenomenology? Most radically, Judith Butler has asked this question a few years ago in her Adorno lectures given in Frankfurt.[3] I here only refer to her thoughts with respect to the question of the self that is accountable for her actions.

Firstly, Butler affirms the 'struggle for recognition' as part of the identity concept, although she sees this rather from the perspective of Foucault's concept of the self. The normative force of the "social" is, in light of the process of self-constitution, based both on a struggle for recognition and on relationships based on social norms that are not even imposed on the self but that the self necessarily appropriates to become a self.

In the tradition of the self-discourse since the 18th century, the teleological perspective and the temporality of the self-relation, or narrative identity and ethical identity, coincide. Quite in line with the Lockean tradition, Butler uses the expression "to give an account of oneself" for this inherent relation between being responsible – in the meaning of accountability – for one's actions on the

---

2   With view to Hegel's concept of recognition, Axel Honneth combines the two traditions, stating the recognition is necessary in personal relationships („love"), in the legal sphere ("rights"), and as societal recognition („solidarity"). Cf. his (Honneth 1992, 1996).
3   Cf. her Adorno lectures (Butler 2003, 2005).

one hand, and narration, on the other. But exactly at the root of the moral self, a concept that is dependent on the notion of an agent who is accountable for her actions, the self must concede that the account of herself is not the account of *her self*. This, however, is not due to the limits of memory as Locke held, but rather due to the social constitution of the self, the origins of which cannot be reconstructed by the self. The self, Butler claims, is sub-jected to norms that she has not chosen but that nevertheless constrain her actions. She is not transparent to herself. She tells her story in dialogue with another person, but depending on whom she tells it, when she tells it, and why she tells it, her "story" will turn out differently. All these stories together both tell and conceal *the self*'s story, which in fact is untellable:

> "If I try to give an account of myself, if I try to make myself recognizable and understandable, then I might begin with a narrative account of my life, but this narrative will be disoriented by what is not mine, or what is not mine alone. And I will, to some degree, have to make myself substitutable in order to make myself recognizable. The narrative authority of the 'I' must give way to the perspective and temporality of a set of norms that contest the singularity of my story."

We can now conclude this section stating that both characteristics of the self are of striking relevance for any moral theory: the dialectic of heteronomy and autonomy that is taken up in the term of recognition, and the self's accountability for her actions, though limited by the constructive dimension of a narrative that reveals and conceals the "self" at the same time.

The Responsible Self

In his book "Oneself as Another", published in 1990, Paul Ricœur has presented a concept of moral identity that takes up the theoretical questions of identity as well as the concept of practical and moral identity. Ricœur holds that the criterion for a successful practical identity can be derived from the self's moral perspective, namely from his or her *aiming at a good life with and for others in just institutions*. Taking up the Aristotelian model of friendship, Ricœur develops the relationship between self and other as symmetrical and as at least partly an act of spontaneous *benevolence* for the other. However, there is also the backside of a spontaneous will, and this is violence, and it is exactly this potential enclosed in human action that renders the moral claim necessary: Only against the background of "evil", Ricœur holds, defined as violence against oneself or the other, arises the necessity to transcend the teleological perspective of "searching the good life". The self must come to acknowledge the deontological claim of morality, which Ricœur articulates in a Kantian reformulation of the categorical imperative:

"Act solely in accordance with the maxim by which you can wish at the same time that what *ought not to be*, namely evil, will indeed not exist." (Ricoeur 1992, 218).

Regarding newer attempts of evolution biology to find the basis of a "moral sense" in benevolence, it is exactly this normative claim that raises problems: While benevolence might be traced in animal behavior, the moral *imperative* to shun violence is a much more difficult normative concept, presupposing the self-reflective stance of the moral self.

This concept of ethical/moral identity, which emerges from the interrelation between care for the self and an interest in living together with others in just institutions, constrained by the recognition of mutual respect and the inhibition and/or prohibition to use (unjustified) violence, describes the moral claim appropriately. It defines morality as a twofold relation of the "good", articulated in the ethical language of desires and goals, and the "right", articulated in the normative language of prescription, namely the "ought" that demands of the (moral) self to respond to the claims of others.

Responsibility, in this approach, is more than accountability for one's action: Care for oneself throughout time and care for the other throughout time enables us to see how memory as *remembrance* must be seen as taking responsibility for the past, how the particular *choices* in the present must be seen as situated freedom and responsibility in the present, and the *effects* of the actions of today must be seen as responsibility for the future.

Following the liberal tradition of reciprocity and mutual recognition, Ricœur grounds morality in the capacity of the self to act self-reflectively, and to care for herself and others, while respecting the others' freedom to act in their own way.

In contrast to Ricœur's interpretation, the Jewish philosopher Emmanuel Levinas attacks the liberal tradition of symmetry and reciprocity as decisive moment of morality, since this is too tightly connected to the "will to survive" and cooperation as condition for survival, as evolution theory has it. Levinas is more Kantian in restricting morality to the normative claim. In contrast to Kant, however, it is not the good will, oriented by the test of the possible universalisation of maxims that found morality. Instead, the *occasion of* and *reason for* morality is a one-sided claim, brought into being by the "face" of the other. Accordingly, and in contrast to Ricœur, Levinas cannot see a "shared perspective" of mutual recognition as the framework of which the normative claim emerges. He distances himself from an ethics that combines care for the self and care for the other and argues instead in favour of the absolute exteriority and alterity of the other. Levinas not only describes the phenomenological relation of self, other, and world by starting with the other, but also anchors his concept of responsibility in the encounter with the other.

In the sensory "event" – which in German may be called a "Widerfahrnis", something that 'befalls' you rather than being initiated by you, the self encounters the other as a kind of "object of the world", as a phenomenon, and yet, as an "embodied counterpart", symbolised in the face, which in Hebrew entails the individuality as well as the exposure of a vulnerable being. It is precisely this vulnerable, mortal face that calls the self to respond:

> "Someone who expresses himself in his nudity – the face – is one to the point of appealing to me, of placing himself under my responsibility: Henceforth, I have to respond for him. ... The other individuates me in the responsibility I have for him" (Levinas 2000, 12).

While the sense of urgency Levinas connects to this responsibility has led many to resist his radical reconfiguration of the self-other-encounter, Levinas himself was convinced that this origin of responsibility must not be regarded as undue, or threatening, or even as a violent intervention into one's freedom and autonomy by the other, but as the "individuation of the self" as moral self. We may debate whether these experiences shape the sense of empathy, or whether the biological disposition to empathy helps to open up for the other in the moral, normative, sense. But I would still argue for a dialectic interaction of biology and morality, insisting on the different languages of empirical explanations, subjective experience, and normative claims. The question that could connect evolution theory or biology and ethics could be stated in Changeux's words:

> "Can a 'new ethics' be devised according to which following Darwin, the propagation of moral norms throughout human societies proceeds through the learning of 'social instincts' of sympathy that have their origin in the evolution of species?" (Changeux/Ricœur 2000, 32)

For Changeux, it is clear that the "brain of a person" maintains the histories of evolution:

> "the evolution of the species, the individual's personal history, and finally the social and cultural history of the community to which the individual belongs... the norms invented by humanity in the course of its history naturally exploit sympathy and the inhibition of violence in the context of a permanent process of cultural evolution." (Changeux/Ricœur 2000, 232).

For Ricœur this argument is precisely what is meant by a historical narrative: The organisation of experiences *retrospectively*, according to the values and normative judgments one holds at the point of narrating.

If responsibility entails the concept of response as well as that of accountability, then the other in deed is the initiator of responsibility, while the agent is accountable for her actions (which could also be the omission to act). Even though the predisposition to this responsible response may be explained in evolution theory, it will not be the "brain" but the moral agent who needs to make choices. I cannot see how neurosciences would change this challenge – the challenge of freedom, unless they would transform the entire concept of humanity that entails the capacity to reflect and make choices.

Evolution theory does not need to go that far; it just needs to be aware of its own constraint of knowledge, namely that the theory of *natural* selection is based upon the hermeneutic principle of *narrative* selection.

Bibliography

Butler, J. (2003): Kritik der ethischen Gewalt. Suhrkamp Verlag, Frankfurt am Main.
Butler, J. (2005): Giving an Account of Oneself. Fordham University Press, New York.
Changeux, P./Ricœur, P. (2000): What makes us think? A neuroscientist and a philosopher argue about ethics, human nature, and the brain. Princeton University Press, Princeton, NJ.
Honneth, A. (1992): Kampf um Anerkennung. Zur moralischen Grammatik sozialer Konflikte. Suhrkamp Verlag, Frankfurt am Main.
Honneth, A. (1996): The struggle for recognition. the moral grammar of social conflicts. The MIT Press, Cambridge, Mass.
Levinas, E. (2000): God, Death, and Time. Stanford University Press, Palo Alto, CA.
Ricœur, P. (1992): Oneself as another. Chicago University Press, Chicago.
Schapp, W. (2004): In Geschichten verstrickt. Zum Sein von Mensch und Ding. Klostermann, Frankfurt am Main.
Sen, A. (2006): Identity and Violence. The Illusion of Destiny. W.W. Norton, New York.

Homo Sapiens, Animal Morabile
A Sketch of
a Philosophical Moral Anthropology

Otfried Höffe

Philosophical ethics is primarily concerned with normative questions which are preceded by meta-ethical reflections. Yet it does not exclude empirical questions, and especially not the question as to how and why natural evolution produced a species, homo sapiens, which is probably the only known species to consist of moral beings. In order to answer this question we must call upon a philosophical discipline currently confined to the fringes of our great debates, a philosophical anthropology, and more specifically a moral anthropology.

The following reflections fall into seven sections. At the beginning I mention the skeptical objections made against such an anthropology (*Skepticism against Anthropology*). They are followed by elements of an argument for a philosophical anthropology (*Concerning Method*) and considerations concerning the biological basis of morality (*The Basic Biological Foundations of Morality*). After these, I insert an excursus on ethical naturalism (*Excursus: Naturalism?*), point out that humans have a non-specified overdrive (*A non-specified overdrive*), ask whether animals can be moral (*Can animals be moral?*) and come to a close with a provisional conclusion (*A provisional conclusion*).

Skepticism against Anthropology

According to a provocative phrase of the philosopher Joseph de Maïstre, "man" is a fiction, a supposed general being that denies the fact that humanity is always embedded in a specific culture: "There is no man in the world. Throughout my life I have seen Frenchmen, Italians and Russians ... but as concerns man, I declare never to have seen him in my life." (see *Considérations* 1884, 74)

Though de Maïstre was a conservative, for some even reactionary intellectual, Marx and Western Marxism, presumably reluctantly, agree with him. Georg Lukács (1922, 204), an early spokesman, highlights the great danger of such an anthropological point of view which "freezes humankind into a rigid objectivity, thereby setting dialectic and history aside". Whereas Lukács had voiced his concern before others had attempted to renewing anthropology, Max Horkheimer (1935) still held to the same critique twelve years later. And a further twenty years later, Habermas (1958/1973, 108) worries in his article "Anthropology" that "an anthropology that operates mostly ontologically, that takes as its object only the recurring and self-same aspects of humanity and human activity, becomes uncritical and in the end turns into a dogmatics with political consequences".

One might, however, wonder quite justifiably whether these objections do not miss the point of the discussion. Protest against rigid forms of objectification had long before been raised and indeed transformed into a recurring theme in philosophical anthropology. The idea that human beings understand themselves differently depending upon their social and historical position, that they have in effect many natures, this idea can perhaps even be traced back to Antiquity and its reception in the Middle Ages. It surely can be discerned in Rousseau's concept of perfectibility (2nd Discourse, Part 1), then in Herder and certainly in anthropology since the 1920s; Sartre (1946) will later say that, because he must make himself into what he is, the human being invents humanity. Long before Sartre – and hence before the critique of Lukács and Horkheimer – Friedrich Nietzsche had expressed this motif in a well-rounded phrase: the human being, he says in the "Genealogy of Morals", is "less defined than any other animal" (Genealogy, 3rd Essay, no. 13).

Concerning Method

How does a philosophical (moral-) anthropology argue? It employs a special kind of hermeneutics as method. It does not interpret specific human cultural elements, but rather empirical findings pertaining to humankind. Hence there are two ways it can proceed which complement one another. On the one hand one can travel into far-off places and search for similarities that people today share with others distant both in space and time. On the other hand one can look close to home and search for that which differentiates humans from other living creatures, especially their evolutionary cousins, the primates. (For more on biological anthropology: Eibl-Eibesfeldt 2004; on comparative primate research: Paul 1998; de Waal 2005; on paleoanthropology: Diamond 1992; also Illies 2006; Brandt 2009).

A moral anthropology that has carefully thought through its methodology does not expect to be able to replace normative ethics. Some empirical sciences, especially young disciplines, do attempt to claim that their own approaches are sufficient. But even their own argumentative logic speaks against the imperialism of some proponents of the empirical sciences. According to the undisputed is-ought fallacy, no ought can be derived from empirical claims, that is, "is-statements". No morality can be founded upon facts alone. A moral anthropology that has carefully thought through its methodology is thus content with the more modest question as to which biological and neurobiological features establish that humans are moral beings. To this attaches the question: Why does morality have a universal basis in (biological) nature and yet is culturally determined? A third question asks: How can morality be understood as a well-founded ought, an imperative, without degenerating into a powerless ought, as Hegelians fear?

Contemporary ethics, as contemporary philosophy in general, has for the most part lost interest in anthropology. A few preliminary remarks are thus in order. Anthropology itself is old. One need only think of Plato, for example in Protagoras (320ff.), and of Aristotle's program of a philosophy of human concerns that encompasses ethics and politics ("Nichomachean Ethics", I 1094b1-11).

The term "anthropology" is however surprisingly new. It is first at the end of the sixteenth century that it appears as a "doctrine of human nature". And later still, essentially in the time of Kant, it becomes an empirical science of humanity. Of especial importance is J.G. Herder's "Treatise on the Origin of Language" (1772/1964). I. Kant also played a significant role, with amongst others his "Anthropology from a Pragmatic Point of View" (1798). In this work he rightly sees that from a methodological point of view anthropology has an intermediate position. It is, namely, a "knowledge of the world" which can be attained neither through philosophy nor through physics, the paradigmatic empirical science of the time ("Anthropology" VIII 3ff.). Later, after a number of specifically moral-anthropological writings from, amongst others, F. Nietzsche in, say, "On the Genealogy of Morals" (1887/1980), philosophical anthropology flourished between the 20s and 40s of the last century, primarily through three philosophers, Max Scheler (1928), Helmut Plessner (1928) and Arnold Gehlen (1940).

Systematically speaking, an anthropology based on biology rests upon the critique of an "idealistic tradition", which takes Aristotle's classic definition of humankind as logos-capable beings and rashly emphasizes the logos side, which it exaggerates into an autonomous intellect. Thus the specific character of the beings is either suppressed or degraded into no more than an expedient and hence heteronomous feature. Starting with Herder and up to Gehlen, anthropology sought a creative alternative to this suppression in a position that went beyond the abstract alternative between autonomous intellect and heteronomous nature. Their attempt to define a new basis for human uniqueness, however, is not entirely convincing.

Inasmuch as this new interpretation does not start out from unity but rather from the other side of the alternative, nature, it is often empirically rich, but thereby at times runs risks turning into naturalism ("biologism"). Interpretatory concepts such as Gehlen's category of instinct-analogy (Instinktanalogie) are overly influenced by the biological concept of survival, and thus fall short of describing humanity's defining features. Yet the new anthropology's basic approach remains plausible. One can certainly explain that which makes human beings unique by first studying the characteristic features of their animal nature and from there move on to language, self-consciousness and reason, and hence to what interests us here: freedom of action and moral abilities.

Against the objection raised by some critics that humankind is defined as an ahistorical being, the new anthropology shows that human biological nature contains a dynamic process that creates not only culture in the singular, those aspects of humanity that reaches beyond the organic and natural, but culture in the plural, its historically differentiated instances. We must therefore abandon the dualistic way of thinking that sees nature here and culture over there. The natural existence of human beings might well be thoroughly shaped by culture, yet these cultural effects are always to some extent determined by organic and natural abilities. At best, anthropology discerns the skeleton of humanity, which then requires cultural and, further, individual features in order to become a tangible being made of "flesh and blood".

Since then anthropology has gained much more experience than was available to its early 18th century and its "classical", 20th century counterparts. But the basic question, and even the two basic answers to it, can already be found in Antiquity: human beings have reason and language and are social, more specifically political beings (see Aristotle Politics I 2, also Höffe ³2006 and Höffe 2001). Of course, human beings, both as individuals and as a species, are initially only predisposed to both. Short of a specific development (evolution) accompanied by individual effort, neither the rational nor the political nature of human beings can become an actuality and flourish.

This requirement that human beings must develop themselves applies also to morality. A human being is not outright an animal morale, a moral being, but an animal morabile, and that in three respects: Human beings are capable of being moral, even called to be so, but must also become so. And so we can now provide a first answer to our third question: Because it requires effort, morality has the character of an ought – and in a more basic sense than usual. Morality does not only appear in the form of a reasoned ought, an imperative. The task of developing morality is itself subject to an imperative, a more basic imperative: Human beings, inasmuch as they are called to be moral, are summoned to grow from beings merely potentially capable of morality into beings actually so.

The Basic Biological Foundations of Morality

Let us turn to the question which addresses the biological basis of morality. Very early on, in myth, one of the precursors of rational thought (see Plato, Protagoras, 320 ff.), two characteristics appear that distinguish human beings from animals; one is a strength, the other a weakness. The weakness: Human organs and instincts are so obviously deficient that it is only slightly exaggerated to speak of the human being as a flawed being. This diagnosis, which one finds now and then, is nevertheless faulty as these deficits find ample compensation in the strength of the human intellect. So our diagnosis concludes: weak body, strong mind. In fact, however, both aspects intertwine, so that weakness turns into a strength: Organs and instincts must not be tuned too carefully to one specific type of environment if intelligence is to have any room to develop. The correct diagnosis should not therefore speak of weak organs and defective instincts, but rather of being open to the world instead of tied to one environment and of a reflexive relation to oneself and to the world instead of a life of immediacy.

What appears at first sight to be a weakness always turns out upon closer inspection to be a strength. We must not underestimate our organs' abilities. Human beings cannot jump as can hares, nor climb like Koala bears, nor swim as can fish, and they certainly cannot fly like a bird. Yet they do possess three of those four abilities, namely running, climbing and swimming, and with the right technology they can even master flying. The newest research indicates our ancestors were such accomplished long-distance runners that they could pursue many an animal until exhaustion; the protein-rich fruit of their hunt then enabled them to evolve larger brains. And in turn this larger brain enabled humans to survive, as they, too, were creatures of prey – and only the smartest prey can outdo their predators. This is indicated by the fact that large predators such as leopards, jaguars and pumas capture more animals with small than with large brains.

Human beings are also very capable in other respects. Human eyes are quite flexible and sensitive to light; some have not only a very fine, but an absolute ear; the tongue can perceive objects that are a fraction of a millimeter thick; and with our hands, we can as well lift heavy boulders as carry out the most delicate work of a goldsmith, a surgeon or a pianist. Human beings are thus generalists who can do just about anything, and hence enjoy through their bodies the benefit of being open to the world. And this enables them, as one example, to live in the most inhospitable regions of the world.

Given the multifaceted openness of their nature, and in particular because of their unbound instincts and their variously useful organs, their intelligence and language abilities allow human beings a second level regulation, that which we traditionally call freedom of action. Human beings are able to establish a relation to the (inner and outer) conditions of their existence, and by means of this reflexive relation can become aware of these conditions, and can name and understand them. They can evaluate the conditions shaping their lives and, depending on the results of the evaluation, can attempt to appropriate these conditions and either creatively rework them or work towards fundamentally altering them. In short: human beings can act in one way or in another, and allow reasons to direct what they finally do or leave undone.

Human beings thus do not live in the present moment alone. They can retrieve experiences from their past; they can learn, experiment and invent in the present; they can use their experiences to anticipate the future and to make plans – though this also implies that they might suffer today of tomorrow's hunger and waste the happiness of the moment by worrying too much about the future.

In order to more accurately define the nature of freedom of action, anthropological findings require a conceptual and linguistic analysis, including metaethical reflections and considerations from the theory of action. One comment here will have to suffice. There are two moments to the relation to self and to reflexivity as concerns action. On the one hand human beings are able to act in a conscious or reflective manner; on the other hand they are able to select among multiple possibilities by recognizing some as their own. Both moments taken together indicate that human beings – regardless of their many handicaps, limits and barriers – are capable of deliberate and willing, and in this sense free, action. Freedom of action should however be understood as a comparative, not an absolute concept. For the moment of deliberation can be more or less correct, clear and complete. There are also various kinds and grades of willingness.

Even in their basic physical needs human beings display their freedom of action. Thus hunger and thirst urge us to eat and drink; but what we eat and drink and when, how often and in which surroundings we do so, how the food and drink are to be found, prepared and stored – all of that is up to us and depends upon further (esthetic, social …) considerations. The urge to take in food and drink is not even necessary in the sense that it must absolutely be met. Be it for matters of health, beauty or asceticism, one can fast for a time, or even refuse any nourishment at all out of religious or political motives.

This multifaceted openness to the world, at which we have only hinted here, reveals a new horizon, a wide though not "specified" field. To use another metaphor, human beings are fields which they must cultivate themselves. This task is conducted along with others, in an ordered and long-term manner, that is, as a culture. Thus human beings are social and cultural beings on account of their biological nature. They might produce various cultures, and in better or worse guises – but they can hardly live without any kind of culture (and its positive morality).

Excursus: Naturalism?

At this point we must ask if one could not circumvent the is-ought fallacy and ground morality (which is part of culture) in biology alone, and therefore with a mere "is". Perhaps we could also circumvent the naturalistic fallacy, the supposed mistake of defining the concept of "good" naturalistically, of defining morality through its usefulness for life.

One side in these debates advocates a biological or ethical naturalism, the positive answer, whereas their opponents prefer the negative answer, ethical anti-naturalism. Both answers usually overlook two aspects. For one, the morality based on the findings detailed above is not a normative or critical, a "moral" morality, but merely a currently valid, positive morality. The is-ought fallacy has not been circumvented: we remain within the purview of the positive, of the "is". For another, this positive morality remains significantly underdetermined. Because we are open to the world, we do require some kind of positive morality; the specifics of it are not, however, determined by this fact. One can understand this lack of determination as a form of tolerance towards various positive forms of morality. And in fact a kind of indifference reigns here that leaves open the possibility, even in some sense the necessity, of the other, the truly moral arguments.

When attempts are made to find a biological basis for the normative concept, they tend to employ criteria such as optimizing the quality of life. This criterion, too, leaves morality underdetermined. There is still a long way to go towards a specific moral program. In particular, the question remains open whether morality should serve the survival of the species (as is apparently often the case in nature) or that of a given culture (as with humans), or even perhaps that of its individual members. We are here confronted with three options – the species, particular cultures or individuals – between which it is difficult to decide on purely biological grounds. There is another alternative, one that is largely foreign to biological considerations: bare life (survival) or a good life. This second option does not simply expand upon the first; it can actually conflict with it.

The animal world admittedly also knows of acts that we would call heroic and altruistic had they been performed by a human being: some individuals do sacrifice themselves for others, in particular females for their offspring. Animals however perform these acts directly, without first asking if they should sacrifice themselves in the first place. This kind of behavior can be extended in two ways, both unknown to the animal world: Should I sacrifice myself for people other than my children, perhaps even for non-relatives? And: Should I remain faithful to a (e.g. religious, political or cultural) conviction even if I must make great sacrifices to do so, perhaps even laying down my life? Even when a positive morality requires sacrifices, genuinely valid moral reasons can be found to refuse such a sacrifice, be it in individual cases or in general. Biology cannot answer such questions, for in the end they require a moral theory.

This also applies to the sophisticated naturalism advanced by the moral philosopher Philippa Foot. According to the thesis indicated in the title of her book, "Natural Goodness" (2001), "good" generally means that which is good for the members of a biological species. As with all animals, even with plants, the good consists also for humans in that which allows them to flourish: "I should prefer to say that virtues play a necessary part in the life of human beings as do stings in the life of bees." (2001, 35)

Foot is in no way claiming that humans are nothing more than bees. She is merely pointing to a semantic and biological similarity. Despite other differences, it is true for human beings just as for animals that "the good is what each species requires in order to live". But she thereby ignores the above-mentioned distinction that animals are primarily concerned only with survival, human beings however with something more. Whereas stings help bees to survive, only one part of morality aims at survival, the rest aims at a good life. Moreover, a conflict arises of which animals know nothing: Should one attempt to simply stay alive and forego a good life, or on the contrary fulfill the obligations of that good life, but at the cost of life itself?

A Non-specified Overdrive

Let us return to the anthropological findings. Among human beings' multiple abilities is a psychological feature, a nonspecific overdrive. The biological basis for this feature lies in a hormone, noradrenalin, that increases performance. Along with intelligence, it does not determine specific ways or goals that humans must pursue – not even the two general goals of individual and collective survival. As such, this overdrive allows such human triumphs as technology and medicine, music, art and architecture, or literature, science and philosophy, but also heroic sacrifices and selfless charity.

These new possibilities are however linked to new kinds of dangers, and human beings can also go in the other direction, turning a biological strength into a weakness. This overdrive enables human beings to strive after more almost without cease: gluttony and sexual indulgence, lust for fame, power and possessions. It is also relevant, morally speaking, that human beings fight for recognition, the consequences of which can be negative such as envy, jealousy and vengeance, but also positive such as forgiveness, sympathy and charity. Humans can also fall prey to delusions of omnipotence. No other animal can desire to "be like

God", and so human beings can be ironically defined as apes who occasionally desire to be God.

It is also apparently only human beings who can, without any inhibition, cause harm to other members of their species, and even to themselves. The great apes are certainly not squeamish, for they will rip out their enemy's fingernails, crush their testicles or tear their throat out. It seems that only human beings, however, are capable of murdering in cold blood, be it on orders, of their own initiative, or sadistically having made the act an end in itself. Here morality lodges an objection. It also justifies this objection and providing criteria for it. The human being is certainly not but a monster, as Sophocles writes. As opposed to apes, only human beings establish friendly relations with their neighbors, carry on commerce and help when catastrophe strikes. Here too morality has its place.

There is an important biological reason for morality: Without biologically pre-programmed checks, human beings must direct their overdrive and learn to operate in a productive rather than a destructive manner. To this end, they must develop into beings who do and leave undone, who live, as they choose.

Can Animals be Moral?

Whether we call it mind, reason or intelligence – the faculty that is responsible for action has an essential cultural component, one which becomes obvious when we consider how much intelligence is dependent upon language. But the reverse also holds true. Culture is to a great extent produced by intelligence. Cognitive science rightly maintains that human brains, being significantly superior to those of apes, permit us to acquire culture, implying in turn that, from an evolutionary point of view, for culture the brain is a dependent variable. We must not hereby forget that an ability to acquire culture implies that there is enormous potential in what individuals can learn themselves as well as great differences in their learning abilities. Adult apes at best reach the intelligence level of 3 to 4 year old human children.

Many animal species have notable mental and social abilities (see Tomasello 1999; de Waal 2005; Perler/Wild 2005). Already the Enlightenment writer d'Holbach (1770/1978, 629, n. 50) warned us not to underestimate the intellectual abilities of animals. And today we know that some animals can categorize the world into objects and events important for their survival. They are capable of remembering a great number of experiences and thereby foresee to some extent future states of their world.

Over the course of their domestication, cats and even more so dogs have learned to interact and communicate in many and subtle ways with human beings, that is, with members of another species. Primates appease one another through gestures or strategically trick one another over and over again. Thus a chimpanzee can pretend to not see a bush in order to pluck its berries alone after the others have moved along. Even birds such as the plover can pretend a broken wing in order to steer a predator away from their nest. A chimpanzee might hide his erection with his hand in the presence of a dominant male, and a female make a conciliatory gesture towards another in order to better snap at her.

Does such chicanery prove that chimpanzees can be moral because they can violate morality, in this case candor and honesty? It is true that these acts of duplicity are not simply goal-oriented, but are also intentional; yet they are only carried out in order to attain the desired goal. The one does not wish to provoke the dominant male, the other is playing a little game. In order to interpret this behavior as intentional acts of deceit, that is, as lying and cheating, we require one more structural element, namely, that the behavior not simply takes place, but does so *as a form of deception.*

A lie requires by definition that one intends to convince someone else of a non-truth. Doing so requires a mental ability that has not been observed even in highly developed animals. The animal must be able to combine in one thought the representation of an actual state of affairs, in this case the acts of concealing or of holding back, with the representation of a merely possible or even imagined situation: that the dominant male is convinced he has no sexual competitors, or that the threatened female is taking her competitor's gesture as a conciliatory one, thereby being duped. Lying requires mental mirroring of the type "I know that you know that I know", in this case: "I believe, that you believe, that I mean the gesture so (as reconciliation ...)".

Controlled experiments show that we can explain these types of behavior as acts that animals have learned to be effective in specific situations. In order to reach their goal, animals do not need to know that their companions are acting on specific convictions. And so we are not here dealing with deception in the strict sense of pulling-the-wool-over-someone's-eyes, and there is hence no moral misdeed to speak of. What some claim to be the beginnings of morality are indeed remarkable phenomena, but they only represent a preliminary stage of morality. The same goes for the ability to remember. A moral subject, what we call a "person", has a specific ability that even primates do not possess (see Markowitsch/Welzer 2005). It is the ability to relate to one's own past and future, what one could call an "autobiographical memory".

A Provisional Conclusion

A moral anthropology that is open to experience also emphasizes other morally relevant viewpoints, such as (as already mentioned above) that human beings care about recognition, even fight for it, which results in such phenomena as envy, jealousy, resentment and revenge, but also forgiveness, sympathy or empathy, compassion, regret and shame. (On the philosophy and psychology of feelings and emotions see Döring 2002; Solomon 2004; Wassmann 2002; on the anthropology of facial expressions see Meuter 2006). Other human innovations are only indirectly morally relevant, e.g. the worlds of technological and medical advances, the worlds of work and play. These are worlds capable of almost limitless progress (perfectibility), and in light of which even the most intelligent chimpanzee populations have been treading water for thousands of years.

From an anthropological point of view, there are two ways morality exists for human beings: intelligence and openness to the world make morality attainable; the dangers associated with humanity's overdrive and openness to the world make it necessary. In any case, human beings are not actually moral from birth as individuals, nor have they been so from the beginning as a species.

As anthropology sees it, morality is a peculiar mix of ought, need and is. Human beings, as open but vulnerable creatures, require commitments, the goodness – and eventually the absolute goodness – of which must be tested through reason. Such tests might not be necessary, but they can hardly be put off forever. And so it turns out that in a first step biology prepares the way for morality. Morality then takes on the specific form of a positive morality within a given culture, and so we needn't view natural endowments and cultural influences opposites. Thanks to a universal human reason this positive morality defers to a critical one, often similar to the first though redesigned.

The biological and neuro-biological nature of human beings thus offer structures within which morality can develop, and which even call it forth in order to survive. It is up to human beings, however, to develop this morality through their own powers and according to their own standards.[1]

---

1 For a closer examination of these issues, see Höffe 2007, primarily chapters 3-5.

Bibliography

Aristotle (1964): Politica. Edited by W.D. Ross. Clarendon Press, Oxford. Engl.: Politics. Translated by B. Jowett. In: J. Barnes (ed.) (1984): The Complete Works of Aristotle. Vol.2. Princeton University Press, Princeton, 1986-2131.
Aristotle (1963): Ethica Nicomachea. Edited by I. Bywater. Oxford University Press, Oxford. Engl. (1985): Nichomachean Ethics. Translated by T. Irwin. Hackett, Indianapolis.
Brandt, R. (2009): Können Tiere denken? Ein Beitrag zur Tierphilosophie. Suhrkamp Verlag, Frankfurt am Main.
Diamond, J.M. (1992): The Third Chimpanzee. The Evolution and Future of the Human Animal. Harper Perennial, New York.
Döring, S. (2002): Die Moralität der Gefühle. In: Deutsche Zeitschrift für Philosophie – Sonderband 4. Akademie Verlag, Berlin.
Eibl-Eibesfeld, I. [3](2004): Die Biologie des menschlichen Verhaltens. Grundriß der Humanethologie. Blank, Munich.
Foot, P. (2001): Natural Goodness. Oxford University Press, Oxford.
Gehlen, A. ([12]1978 [1940]): Der Mensch. Seine Natur und seine Stellung in der Welt. Athenaion, Wiesbaden. Engl. (1988): Man. His Nature and Place in the World. Columbia University Press, New York.
Habermas, J. (1958): Philosophische Anthropologie. In: Diemer A./ Frenzel I. (eds.): Fischer Lexikon Philosophie. Fischer, Frankfurt am Main. Reprinted in: Habermas, J. (1973): Kultur und Kritik. Suhrkamp Verlag, Frankfurt am Main, 89-111.
Herder, J.G. (1979 [1772]): Abhandlung über den Ursprung der Sprache. Edited by H.D. Irmscher. Reclam, Stuttgart. Engl.: Treatise on the Origin of Language. In: Forster, M.N. (ed.) (2002): Philosophical Writings. Cambridge University Press, Cambridge, 65-166.
Höffe, O. [3](2006): Aristoteles. Beck, Munich. Engl. (2003): Aristotle. State University of New York (SUNY), New York.
Höffe, O. (2001): Aristoteles' Politische Anthropologie. Reihe: Klassiker Auslegen Bd. 23, Aristoteles. Politik. Akademie Verlag, Berlin, 21-35.
Höffe, O. (2007): Lebenskunst und Moral. Oder Macht Tugend glücklich? Beck, Munich. (paperback 2009). Engl. (2010): Can Virtue Make Us Happy? The Art of Living and Morality. Northwestern University Press, Evanston.
Holbach, P.-H. Th. d' (1770): Système de la nature ou des Loix du monde physique et du monde morale. London. Engl. (1970): System of Nature. Translated by H.D. Robinson. Burt Franklin, New York.

Horkheimer, M. (1935): Bemerkungen zur philosophischen Anthropologie. In: Horkheimer, M. (1988): Gesammelte Schriften. Bd. 3. Fischer, Frankfurt am Main, 249-276.
Illies, C. (2006): Philosophische Anthropologie im biologischen Zeitalter. Zur Konvergenz von Moral und Natur. Suhrkamp Verlag, Frankfurt am Main.
Kant, I. (1968 [1798]): Anthropologie in pragmatischer Hinsicht. In: Kants Werke. Akademie Ausgabe, Bd. VII. De Gruyter, Berlin, 117-334. Engl. (1974): Anthropology from a Pragmatic Point of View. Translated by M. Gregor. Martinus Nijhoff, The Hague.
Lukács, G. (1967 [1922]): Geschichte und Klassenbewußtsein. Studien über marxistische Dialektik. Photomechanical Reproduction, Amsterdam.
Maïstre, J. de (1814): Considérations sur la France. In: Maïstre, J. de (1884): Œuvres complètes. Nouvelle Edition, Bd. ½. Lyon, 1-184. Engl. (1974): Considerations on France. Translated by Richard A. Lebrun. McGill-Queen's University Press, Montreal.
Markowitsch, H.J./ Welzer, H. (2005): Das autobiographische Gedächtnis. Hirnorganische Grundlagen und biosoziale Entwicklung. Klett-Cotta, Stuttgart.
Meuter, N. (2006): Anthropologie des Ausdrucks. Die Expressivität des Menschen zwischen Natur und Kultur. Fink Wilhelm, Munich.
Nietzsche, F. (1980 [1887]): Zur Genealogie der Moral. München. In: Kritische Studienausgabe. Edited by G. Colli and M. Montinari, dtv, Munich, vol. 5, 246-412. Engl. (1998): On the Genealogy of Morality. Translated with an Introduction and Notes by M. Clark and A. J. Swensen. Hackett, Indianapolis.
Paul, A. (1998): Von Affen und Menschen. Verhaltensbiologie der Primaten. Wissenschaftliche Buchgesellschaft, Darmstadt.
Perler, D./Wild, M. (eds.) (2005): Der Geist der Tiere. Philosophische Texte zu einer aktuellen Diskussion. Suhrkamp Verlag, Frankfurt am Main.
Platon: Protagoras. In: Duke, E.A. (ed.) (1995): Platonis Opera. Vol. 1. Oxford University Press, New York ff., 309-362. Engl.: Protagoras. In: Hamilton, E./ Cairns, H. (eds.) (1961): The Collected Dialogues of Plato. Princeton University Press, Princeton, 308-352.
Plessner, H. (1981 [1928]): Die Stufen des Organischen und der Mensch: Einleitung in die philosophische Anthropologie. Gesammelte Schriften Vol.4., Suhrkamp Verlag, Frankfurt am Main.
Rousseau, J.J. (1933): Discours sur l'origine de l'inégalité parmi les hommes. In: The Social Contract and the Discourses. Translated by G.D.H. Cole, revised and augmented by J.H. Brumfitt and J.C. Hall, 1993, A. A. Knopf, New York.
Sartre, J.P. (1946): L'existentialisme est-il un humanisme? Paris. Engl. (1963): Existentialism and Humanism. Translated by P. Mairet. Methuen, London.

Scheler, M. (^{16}2005 [1928]): Die Stellung des Menschen im Kosmos. Bouvier Verlag, Bonn. Engl. (2008): The Human Place in the Cosmos. Translated by M. Frings. North Western University Press, Evanston.
Solomon, R.C. (ed.) (2004): Thinking about Feeling. Contemporary Philosophers on Emotion. Oxford University Press, Oxford.
Tomasello, M. (1999): The Cultural Origins of Human Cognition. Harvard University Press, Cambridge.
Waal, F. de (2005): Our Inner Ape. A Leading Primatologist Explains Why We Are Who We Are. Riverhead Books, New York.
Wassmann, C. (2002): Die Macht der Emotionen. Wie Gefühle unser Denken und Handeln beeinflussen. Wissenschaftliche Buchgesellschaft, Darmstadt.

Enhancement and Evolution[1]

Sarah Chan

Humanity is constantly reinventing itself. From the earliest days of our species, when one of our ancestors picked up a burning stick and kindled it into a fire, to the present day; humans have been altering their environment and shaping the world around them. The history of our species is a stream of discoveries – major and minor – which have allowed us to progress and direct, to some extent, the course of our evolution.

Probably the most important technological advances in recent history have been those produced by biomedical science. Our ever-increasing understanding of human biology, and the consequent ability to treat and prevent disease, have improved healthcare phenomenally during the past century. The emerging medical technologies suggest even greater possibilities for the relatively near future such as stem-cell therapy and regenerative medicine, genetic manipulation and new pharmacological agents.

But why should we limit ourselves only to treating disease and injury? These same technologies actually hold the potential to allow us to improve further upon myriad aspects of human function – that is, to enhance current and future generations. This possibility has raised considerable debate: Is human enhancement acceptable and how far should we go in pursuit of this goal? What will the use of enhancement technologies mean for the future of humanity, and what exactly do we mean when we speak of humanity?

This paper has three parts. In the first part I set out an argument in support of human enhancement, covering in brief the main issues that have arisen during the debate; and argue that enhancement is not only morally acceptable but is a morally worthwhile goal to the point of being an obligation. In the second part, I consider the implications of human enhancement (including but not limited to

---

1 This paper, delivered at the international conference Evolution and the Future, Belgrade, 14-17 October 2009, is partly based on previously published research including the following papers: Sarah Chan, Humanity 2.0? Enhancement, evolution and the possible futures of humanity EMBO Rep. 2008; 9 (Suppl 1): S70-4; Sarah Chan and John Harris: In Support of Human Enhancement Studies. In: Ethics, Law and Technology 2007; 1(1): Art 10; Sarah Chan: Should we enhance animals? In: Journal of Medical Ethics 2009; 35(11): 678-683.
The author wishes to acknowledge the stimulus and support of the Wellcome Trust Strategic Programme on The Human Body: Its Scope, Limits and Future.

genetic enhancement) in the context of our future evolution. Thus far the argument follows along somewhat well-rehearsed lines developed in the context of human enhancement. In the third part of this paper, however, I aim to explore one aspect of the argument regarding enhancement and evolution that has received less attention so far: the implications for non-human evolution.

In Support of Human Enhancement

So, what is enhancement? I propose that human enhancement is anything that improves our function: any intervention that increases our general abilities and allows the individual to flourish. In other words, an enhancement is something that is of benefit to the individual – as John Harris puts it: "If it wasn't good for you, it wouldn't be enhancement" (Harris 2007). Let us exclude from the present consideration those procedures often termed "enhancements" that are of dubious overall benefit (for example breast or penis augmentation, or the taking of anabolic steroids to increase muscle mass); equally we are not talking of "designer" modifications which are more akin to aesthetic or fashion preferences than to improvements: hair colour, eye colour, or physique. An enhancement in the sense that I am using the term is something of benefit to the individual.

If we accept this, it follows that there can be nothing morally wrong with human enhancement per se: we already accept and actively encourage the use of enhancements in various aspects of our activities. Dietary supplements to improve health and wellbeing; prosthetic limbs for the disabled; vaccination to increase immunity to disease, reading glasses, opera glasses and hearing aids : all of these represent enhancements. It may be argued that some of these constitute medical treatment rather than enhancement, but is this a meaningful distinction?

It is difficult to see how there could be a moral difference between improving reduced function to normal levels and improving normal function to supernormal. Consider: the very concept of what we think of as normal has changed across time, through altered environmental and genetic factors. In a health context, better nutrition, scientific knowledge and the intervention of modern medicine have all produced improvements to the "normal" condition. (For example, it was once considered normal for women to die in childbirth and the pain of childbirth was considered part of the authentic experience). Proponents of the notion of "species-typical function" as a benchmark for distinguishing between treatment and enhancement often ignore the fact that many interventions considered to be treatment actually alter typical function, and that the present definition of "normal" includes many species-atypical features. At an individual level, therefore, the concept of normalcy lacks both precision and moral content; "treatment" and "enhancement" are morally indistinguishable. It is true that the fact that there are many "shades of grey" here does not negate the existence of black and white; similarly the fact that we cannot draw a bright line between therapy and enhancement does not mean that some things are not clearly therapy, others clearly enhancement. But the question we should be asking ourselves is not whether we *can* draw a line but whether that line has any moral significance, and I submit that it does not.

Those who believe that human enhancement contradicts our moral values will, therefore, either have to show that the current benefits of modern medicine and science – which they call therapies – are somehow implicitly different to those that we might develop in the future – which they call enhancements – or reject all the enhancements that we now enjoy, on pain of moral hypocrisy.

A common but fundamentally absurd objection to genetic enhancement is that it is somehow "unnatural". Aside from the fact that it is difficult to distinguish what, exactly, constitutes "natural", there is no logical or ethical reason why that which is natural should be preferred to that which is unnatural. If humans are morally obliged to be inert, passive players in the game of life, to refrain from exerting any influence or control on the world around us, then a vast proportion of human achievement to date must be classed as unethical, since it is unnatural. Much of human activity is against the course of nature: indeed, it may be argued that our capacity to shape our environment according to our conscious desires and wishes is a crucial part of what it is to be human.

Arguments against enhancement often single out genetic enhancement in particular as being unacceptable. If enhancement itself is not wrong, is there anything that is added by a genetic component that makes a difference? There is no special moral significance to genetics itself (although genetics has acquired a sort of popular mystique); genetic technologies are factually different but morally similar to enhancements we already utilise. Genetic manipulation might be perceived as less acceptable because it is a relatively new procedure – at least when applied to human beings – and therefore carries the possibility of unforeseen risks. Gene therapy, in particular, has received bad press after a few cases in which potential dangers did manifest. At the current level of technology, this is probably valid: our grasp of genetic modification and its effects remains tenuous, and to apply these techniques as they exist at present would involve a substantial and probably unacceptable chance of unpredictable, possibly harmful consequences. It does not, however, follow that it would be wrong to develop the technology to a point where it is safe and subsequently apply it. Although the risks may never be eliminated entirely, the potential benefits are sufficiently significant to justify incurring a certain amount of risk, as is the case with many activities (enhancing or otherwise) that we willingly undertake. It is hard to think of a medical advance that has not been more risky in the early stages than has later proved to be the case.

So is there anything intrinsically wrong with genetic manipulation? There certainly cannot be a moral proscription against modifying the somatic genome: we are at liberty to make changes to our physical bodies in terms of their appearance, condition or health, and DNA is simply a part of that – in fact, somatic mutation occurs all the time. Arguments against human genetic modification have therefore concentrated in particular on the ethical unacceptability of germ-line gene therapy. The distinguishing feature of germline genetic modification – as opposed to somatic genetic modification – is that any changes made to the genome will be heritable and therefore affect not only the individuals treated, but also their descendants – and by extension, the future human race as a whole.

The hereditary nature of germline genetic modification may give rise to objections regarding the consequences for future generations: that we do not have the right to decide what genetic heritage they should receive.

Yet children can never exert a choice over the genes that they are born with, regardless of whether their parents do. Those who believe in genetic determinism must realize that abandoning our children to the mercy of the 'genetic lottery' does not free them from the tyranny of their genes, rather, it merely removes any element of choice on the part of anyone as to what those genes are. Those of us, by contrast, who look beyond genetic determinism to see that it is far more than genes that determine the future of our children, will probably also realize that – to the extent that genes do determine some aspects of our lives, in particular our health and associated quality of life – it is surely far better to have a predisposition to a healthy life than to risk a higher chance of suffering and disease.

The corollary to the above argument is that parents themselves should not seek to make such choices and that attempting to enhance children represents what Michael Sandel has referred to as "a kind of hyperagency – a Promethean aspiration to remake nature, including human nature, to serve our purposes and satisfy our desires" (Sandel 2004, Sandel 2007). According to this argument, parental virtue requires acceptance rather than control and an "openness to the unbidden", rather than the "hubris [of] an excess of mastery" – that is, an appreciation of the 'gifted' nature of human achievement, rather than an aspiration to increase these achievements. Choosing to enhance is therefore wrong both because it violates the principles of good parenting and because of the negative social consequences that will follow the abandonment of such principles.

In each case, however, it is not clear whether the availability or use of genetic enhancements will result in the predicted disintegration of social values, or if allowing parents to make choices about their children contravenes the requirements of parental responsibility. It is true that the use of genetic enhancement technologies will involve parents making decisions about the genetic heritage of their children. However, once the technology exists and has been proven to be safe, to refrain from using it is likewise to make a decision about the genetic inheritance of our descendants – specifically: that they will not receive such enhancements and will be denied the consequent benefits. If enhancements are (as suggested above) beneficial, all things considered, it is unlikely future generations will thank us for refusing to adopt them. Abdicating choice is not the action of a responsible parent; exercising choice wisely is.

Of course, this applies equally to both genetic and non-genetic interventions. Every parent wants to have a healthy child and there is nothing wrong with this. Indeed, we might look askance at a parent who did not claim to want the best for their children. Yet what do we mean by healthy? The only logical answer to this question is "as healthy as possible". For example, we know that certain maternal behaviours during pregnancy can have adverse consequences for child health: alcohol and tobacco use increase the risk of low birth weight and associated developmental delay, and poor maternal diet can increase the chances of developing Type 2 diabetes in later life. We encourage mothers-to-be to improve their child's health by avoiding these factors; indeed, we consider it irresponsible for them to fail to do so. Similarly, we advocate folate supplements as a positive measure to decrease the risk of spina bifida. That is, we see it as a strong obligation on parents to use the available knowledge and medical technology to maximise their children's health.

Surely, then, this argument also applies to genetic enhancement. If we could, with safety and certainty, engineer immunity to viral infection, protection from heart disease or reduced susceptibility to cancer in our children, we should do so. In other words, not only is genetic enhancement morally acceptable, but if and when it becomes safe and affordable, there will be a moral imperative to use it for the benefit of future generations.

Enhancement and Human Evolution

What does human enhancement mean in terms of human evolution? A prominent concern about enhancement technologies is that their use might compromise our humanity and transform us into something other than – and perhaps beyond – human, thereby jeopardizing our species. These concerns, I would argue, are misplaced; but an examination of the potential impact of enhancement technology on humans and humanity might lead us to view enhancement in a different light and help to define what it means to be human. In fact, the use of enhancements – genetic or otherwise – will not cause us to cease to be human or to lose the essential qualities of humanity; indeed, enhancements and the desire to use them might be seen as an expression of our humanity.

To understand this, let us examine some of the concerns regarding enhancement and the future of humanity.

These concerns include worries about the loss of genetic diversity: that we will become a race of clones – either literally, through the use of reproductive cloning to create armies of identical beings, or because our uniform desire for enhancements will result in overall genetic uniformity. Others seem to have the opposite fear: that changing our genome and creating genetic differences will lead to the fragmentation and destruction of the human race. George Annas, for example, has described genetic engineers as "potential bioterrorists" because of the possibility of genocide based on engineered genetic differences (Annas 2001). An early draft of the United Nations Educational, Scientific and Cultural Organization (UNESCO) Universal Declaration on the Human Genome states that, "the human genome must be preserved as the common heritage of humanity" (UNESCO 1997). However, the only way to eradicate all genetic differences that might lead to genocide, and the only way to preserve the human genome exactly as it is today, is universal human reproductive cloning – a course of action that neither Annas nor UNESCO would be likely to favour. These arguments, then, are confusing and contradictory, yet they strike a note of concern with many. Even if this concern is misplaced, or misdirected, we might well wonder what the consequences of enhancement will be for humanity as a whole.

We are familiar with the idea of Darwinian evolution: natural selection acting on genetic variation to produce long-term changes in our genetic make-up. There is also a sense in which human society and culture can be said to undergo evolution, as ideas that are more successful – whether because they are correct or intuitively attractive or both – will propagate themselves: a process known as meme evolution. This, of course, includes the progression of technological ideas.

It is reasonable to say that the relationship between humans and technology has probably influenced the course of our genetic and cultural evolution in the past. For example, it seems likely that the technological developments since prehistory have changed the selection pressures acting on human beings – one can plausibly suppose, for example, that the transition from hunter – gatherer to agricultural societies would have altered the environment to allow different genes to be favoured. Moreover, technology has undeniably influenced the course of social development.

It is often said, with reference to the human race, that we have halted evolution by doing away with natural selection. It is true that modern medicine allows many individuals to survive and reproduce when they might not otherwise have done so; yet, we must remember that the marked reductions in infant mortality, deaths from disease and so on are only a product of the past century or so – a mere few generations. In the context of the roughly three billion year history of life on Earth, it seems to be a little premature to declare victory over natural selection – although we are probably the first and only species that understands evolution and might be in a position to influence its course deliberately.

Even if we were to remove the influence of natural selection entirely, it would not necessarily mean the end of evolution – if by evolution we mean social change or cultural evolution – or the end of genetic change. In fact, although the collective human genome is not changing appreciably faster than it has in the past, the development of human society as a whole – that is, evolution in the cultural sense – is progressing faster than ever, and is constantly accelerating. What we want to do and what we dream of doing is rapidly outpacing what we can do with the genomes and the bodies that we are presently born with. Hans Moravec, a robotics expert, has expressed this as, "the drag of the flesh on the spirit [...] the problem is that cultural development proceeds much faster than biological evolution" (Moravec 1989). Interestingly, he noted this in the context of artificial intelligence and wondered whether machines might overcome the "drag of the flesh" from which humans suffer. With the new biological technologies offered to us, we might be able to overcome the drag in our own bodies.

Modifying the human genome might be one way of doing this. Some see it as tampering with the process of natural evolution that produced the human genome in its current form. True enough, our present genome is, by definition, a successful one: it is still around today. However, evolution so far has operated somewhat in the dark. The happenstance of natural evolution has also bequeathed us a legacy of genetic failings: susceptibility to disease, cancer and the depredations of old age, to name but a few. We are, as Richard Dawkins might put it, stranded on our own particular peak of 'Mount Improbable', lumbered with all the genetic encumbrances of our evolutionary history, good and bad. We have now, almost within our reach, the power to transcend this situation: to change our genotypes, much as we can already change our phenotypes to improve health and quality of life. There might be no guarantee of success; however, in comparison to natural evolution, any genetic modifications aimed at improving the human condition will at least be evidence-based rather than random, and unless blind watchmaking is to be the new gold standard in scientific research, that is as good as it gets with current technology.

This, then, is the future of human enhancement and human evolution – a process that John Harris has dubbed "enhancement evolution", the next phase for humanity, in which we control our future and our lives in a way that extends not just to our genes but to our physical bodies and our minds. The possibility of radical changes that this might one day bring about in us has also led Harris to say that "in the future there will be no more humans" – he, of course, thinks that this is not something to worry about.

Enhancement technologies, provided their application is correctly managed, will ultimately be of benefit to humankind and, as such, we have good moral reasons to pursue them. They will not, as some fear, destroy humanity by turning us into something other – something more – than what we are – because we have always wanted – and will probably always want – to be more than we are. Yet by the same token, we will always remain 'what we are' – in the sense of that which really matters to us, that which makes us human. Surely, being human and the definition of 'humanness' should not depend arbitrarily on a particular combination of abilities and limitations – to run no faster than 25 miles per hour, to live no longer than a century. Indeed, logically it cannot depend on such limitations, or we would be forced to admit that we ourselves are no longer human compared to our predecessors who had more limited abilities and life spans. Instead, what make us human are our aspirations, our awareness of ourselves as beings in the world (including our limitations), our ability for self-contemplation and reflection, and the desire to attempt to change what we see. We are not designed to remain "passive, inert players in the game of life" (Chan/Harris 2007). What makes us uniquely human is the ability to shape our own destinies according to our desires – and genetic and other enhancement technologies provide further means for us to do so. In this sense, enhancements and the desire to avail ourselves of them are an expression of our essential humanity. The advent of new forms of human enhancement on our technological horizon does not therefore signify, as some have warned, the end of humanity. Rather, it is just the next step in the continuing process of human evolution. The kind of beings that exist in our future may no longer be *Homo sapiens*, but still I think we would recognise them as us.

Enhancement Evolution: The Next Phase

The question is then "Where do we want to direct our evolution?", perhaps even where should – given that we *are* moral animals who can understand and reflect on the concept of should – we direct our evolution? In other words, having asked and answered the question of what role enhancement might play in evolution, let us now turn the question on its head and ask, what aspects of evolution do we want to enhance?

We have discussed at some length the possibility of physiological, cognitive and genetic enhancements. Another possibility that deserves further consideration is that of moral enhancement. Having evolved to be moral animals, creatures with the capacity for moral reasoning and decision-making, what do we then do with that capacity? Can we or should we use our morality to enhance both our moral natures and our capacity for morality? Most interestingly, in the context of our definition of enhancements as something of benefit to the individual, are improvements to one's moral nature really enhancements? Is it always better for us to be more moral?

These are fascinating questions worthy of extensive consideration, but beyond the immediate scope of this paper. The question that begs immediate consideration, having decided that if we were to evolve further, through genetic and other changes, we would still be essentially "human" in the sense of being "one of us", is why we should limit the application of enhancement evolution to "humans. In previous papers at this meeting, it has been suggested that we should extend the traditional boundaries of ethics from solely humans to including the non-human world. With respect to enhancement technologies and evolution, this leads us to ask: Should we enhance animals?

This may seem like a problem of science-fiction rather than ethics, but the practical implications of enhancement technologies make it imperative to consider the ethical questions involved, while the answers obtained may have unexpected implications for our attitudes towards human enhancement.

To frame the question, let us consider the following hypothetical scenario:

Consider a situation where you knew your child was going to be born with a congenital brain condition such that his cognitive abilities would be restricted to a level equivalent to that of a higher primate, such as a chimpanzee. If under these circumstances you had the option of treatment to restore your child's intelligence to a level approaching normal, it would be extremely remiss of you (to say the least) to decline it!

Now consider an alternate situation. A baby is going to be born with a congenital condition limiting his native cognitive abilities to that of a chimpanzee. The condition is that the baby is, genetically and biologically, a chimpanzee! Why do we feel that in this case there is no obligation to provide treatment or enhancement to increase his intelligence to a level approaching humans?

There are numerous possible answers to this question, some of which may be valid, others not; but we will return to these later. Meanwhile, why should we discuss animal enhancement? Aside from being an interesting ethical question, it is also a practical one: enhancing animals is something we already do. The sorts of technologies that, it is suggested, could be used for human enhancement will in all likelihood be first tested on animals; and indeed some of them are already in use on animals – for example, genetic manipulation and drug testing. A consideration of animal enhancements might also, as I have suggested, cast new light on the discourse over human enhancement, in that what we learn about the reasons to enhance or not to enhance animals could extend to our understanding of why we should or should not enhance humans. Finally, in this exploration I propose that there might be some positive moral reasons to enhance animals (as it is suggested there are to enhance humans).

A discussion of the ethics of "animal enhancement" first requires us to consider what is meant by the term. In one sense, humans have already been "enhancing" animals for many centuries. Generations of selective breeding to intensify certain traits in domesticated species have resulted in strains of animals that have been enhanced in the sense that some aspect of their function has been increased. This "enhancement" has resulted in dairy cows that produce more milk, horses that are larger and stronger, and hundreds of dog breeds of different shapes and sizes. Moreover, genetic manipulation technologies now allow us to effect genetic changes more quickly, effectively and directly than by traditional breeding methods; and as well as make use of existing genetic variation, introduce new genes in our quest to transform existing species. Transgenic mice bearing introduced genes from humans or other species are commonly used in biomedical research (see for example Galas 2003; Benveniste 2007) another example is ornamental fish genetically modified for fluorescent colouration (Knight 2003).

Technologies for human enhancement could also be applied to non-human animals – indeed it is probable that many enhancement technologies, if they have not already been developed and tested on animals, will be so before they are first applied to humans. Examples of these "animal-enhancing" technologies include genetic manipulation – for example, *C. elegans* (Collins 2006), *Drosophila* (Rose 1989) and mice (Lithgow 2000) are among the types of experimental organisms that have been genetically engineered for longevity. Genetic modification has also succeeded in producing a "super"-mouse with quadrupled muscle mass (Lee 2001; Lee 2007), suggesting a possible pathway to physical enhancement for human athletes. Another potentially enhancing technology is the creation of human-animal chimeras. Mice with humanised immune systems have been created as experimental organisms (Shultz 2007); perhaps of more concern are human-animal neural chimeras such as the human neuron mouse (Tamaki 2002; Muotri 2005; Greely 2007), or primates with human neural stem cells in the forebrain (Ourednik 2001) – although the latter have so far been developed only to embryo stage.

Arguments relating to what might be termed animal enhancement have largely focused on the application of these various technologies. They tend to fall into the following groups:

Risk-based concerns
Dislike of the unnatural
Concern for animal welfare
Concern over making animals more human

Ethical discourse over human enhancement has moved beyond arguing about the technology to a more general consideration of the principles of enhancing humans. Similarly a proper ethical discussion of animal enhancement requires us to develop a definition of the term.

The obvious question when it comes to classifying animal enhancements is: enhancements from whose point of view? Considering the range of technological interventions that might be termed "animal enhancement", it is possible to identify at least three definitions. An enhancement of an animal may be something that:

1) Produces an increase in some natural function or confers a novel function
2) Improves some aspect of the animal for human purposes
3) Enables greater fulfilment of the animal's own interests

These categories will obviously overlap in some cases, though not all. An increase in some natural ability may be in an animal's interests as well as serving human purposes. Conversely, many "enhancements" that increase the animal's fitness for human purposes might be considered to be against the animal's interests – for example, highly specialised dog breeds that suffer from health problems as a result of selective breeding for a particular appearance.

What can we say about these possible definitions from an ethical point of view?

In relation to the first definition (enhancements of natural or novel function), we have argued in considerations of human enhancement that there is no special moral quality associated with the natural, nor any moral proscription against the unnatural. It is therefore implausible that either increasing a natural function or introducing a novel function should be morally right or wrong in itself. Yet it is easy to imagine potential consequences of these modifications that might be good or bad from the perspectives of various beings concerned. A cow whose resistance to disease is enhanced, whether by increasing its "natural" resistance or introducing novel genes, will probably live a better, disease-free life. On the other hand, genetically engineering pigs to serve as immunocompatible organ donors for humans is obviously good for the human recipients, but might be bad for the pig if the premature ending of its life as an organ donor could be said to harm it.

If not wrong in itself, therefore, animal enhancement must be right or wrong consequentially, in relation to the sorts of entities that matter morally. This brings us to the second definition: enhancement for human purposes or benefits. Humans matter morally, therefore an animal enhancement which produces a benefit to humans is a good thing, at least in terms of those humans. This view of the utility of animal enhancement cannot be ignored.

We might object to the use of animals to benefit humans on the grounds that it constitutes a sort of instrumentalisation. The Kantian imperative against instrumentalisation, however, is only relevant in terms of beings of a certain moral status Kant thought animals didn't qualify – if they do not, then they are not harmed by instrumentalisation. But they can still be harmed or indeed benefited directly by enhancement. This brings us to the third definition, and the one that I suggest is most appropriate as a definition of animal enhancement: an enhancement in animal's interests. If there are moral reasons for or against animal enhancement they must derive at least in part from the moral status of animals and our relationship to them.

If we are to accept the idea of animal enhancement being an intervention that is in the animals' interests as an appropriate framework for considering the ethics of animal enhancement, we must establish a number of points. Are animals proper subjects of moral consideration: do they matter morally? If not, their interests will have no bearing on our ethical analysis. And what are animals' interests? If we are to say whether or not an enhancement is in an animal's interest and hence ethical, we will need an adequate account of what these interests are.

There has been much work done on the moral status of animals, but here is not the place to recapitulate it in detail. Instead, we can start from the following premises:

Animals can have interests.

These interests vary depending on the animal's capacities: for example, sentience gives rise to an interest in not feeling pain; the capacity for desire leads to an interest in fulfilment of those desires; and for animals who are self-aware, there may be an interest in continued life.

If animals can have interests, they can be harmed or benefited in view of those interests.

We have a moral obligation to animals, in virtue of their interests, to benefit them and not to harm them

Although each of these premises may be subject to philosophical scrutiny and criticism, most accounts of animal ethics do accept some version of these and the argument that follows from them, for example to create the moral obligation not to cause animals to suffer (unduly or unnecessarily).

In terms of animal enhancement, this means that we have an obligation to enhance animals where it is in those animals' interests to be enhanced, and to refrain from doing so when it is against their interests. Therefore technological interventions affecting animals will be morally permissible if they do not cause harm to any creature's interests, and may be morally obligatory when they also serve the interests of the enhanced animal or other animals, including humans.

What are the implications of this conclusion in terms of animal enhancement? If we have obligations to act in animals' interests, to benefit them and not to harm them, then we have an obligation to use enhancement technologies on animals when it is in those animals' interests, and to refrain from doing so when it is against their interests. The greater the interest, the stronger the obligation in each case. We may also have obligations to animals (including humans) to enhance other animals when it benefits the first group of animals to do so; but these obligations will have to be considered in conjunction with whether the enhancement is in the interests of the enhanced animals themselves and the strength of our obligations in this regard.

How we determine in practice whether a given intervention is in an individual animal's interests – that is, whether or not it constitutes an enhancement in the beneficial – may sometimes pose difficulty: how does one know what is best for others? But this problem also arises, although to a lesser degree, in assessing the benefits of some forms of human "enhancement". Although we might concur in general that a life free from the fears of cancer or the infirmities of age is (all other things being equal) probably better than one burdened by them, we might not agree so universally that having, for example, synaesthetic capabilities or the ability to sense electromagnetic fields would necessarily be good for us. If we accept the interest-relative definition of enhancement, whether human or animal, then it follows that whether or not something constitutes enhancement must ultimately be a subjective question: one man's meat (or one mouse's cheese) may be another's poison! We must remember, however, that unless there is a good reason not to have a modification, even if the individual does not view it as enhancing, then it is at worst neutral rather than negative.

Humans can often decide for themselves whether something benefits them or not; but when this is not the case (for example when making decisions for children regarding potential enhancement) we must judge what is most likely to be in the individual's interests. When it comes to making decisions for others regarding enhancement, then, it is a matter of assessing likely rational preferences as to whether something is likely to be of benefit to the individual concerned.

We are often able to make reasonable assumptions about what is good for human individuals and for humans in general because, as humans ourselves, we are inside the community of judgement: we share many experiences, have many features of our lives in common, and evidence suggests to us (even if it can never be conclusively demonstrated) that we may have similar subjective experiences and inner lives. Even this is not foolproof, as the vast body of debate in medical jurisprudence over "best interests" demonstrates. With respect to non-human animals, though, we are at a further disadvantage when it comes to making substituted judgement decisions, because of the many aspects of our lives and experience we do not share with them.

Perhaps this is a matter of degree rather than substance: it is easier for us to determine what will be good for a human whose lifestyle, age, cultural background, abilities, education, life experiences and so forth are similar to our own, than to determine what will be good for someone in circumstances greatly different to ours. Likewise, the differences between human and animal lives create a greater but not (or at least not always) insurmountable barrier; benefits that are relevant to a specific interest may be generalisable across all individuals of whatever species that share that interest – for example, any creature that has an interest in continued life will presumably benefit from an increase in the length of that life. It is less clear how to deal with interests that may be species specific, although we may often be able to hazard a guess: humans as a species lack supersonic hearing, but we might suppose that a dolphin without sonar abilities would probably have a worse life than one with.

We can say in general, however, that technological interventions affecting animals will be morally permissible if they do not cause harm to any creature's interests, and may be morally obligatory when they also serve the interests of the enhanced animal or other animals, including humans. Our obligations to enhancement are likely to be greatest when we are considering enhancements that will probably cause harm to none, will be in the overall interest of either the enhanced creatures or other moral subjects, and will be in the interests of creatures with higher psychological capacities such as self-aware entities or persons.

The existence of such obligations towards non-human animals does not negate the existence of other obligations we may have, not least towards humans. Like all obligations, our obligations towards non-human animals in respect of their interests do not bind absolutely but must be viewed in the context of the complete network of obligations to which we are subject in respect of all creatures within our moral sphere. As our obligation to refrain from harming others may yield to our need to protect ourselves in the case of self-defence, the obligations we hold towards animals must be weighed against other, potentially conflicting obligations to determine the optimal moral course of action.

Therefore, if I say that we have an obligation to enhance animals in their interests, I do not mean that this holds even at the cost of detriment to other creatures such as humans and neglecting our obligations in this regard; just that the obligation exists to be weighed against these other possible obligations.

How we balance the scales to determine which, out of all the competing obligations we might have towards other creatures (human and non-human alike), are the most pressing and thus dictate our optimally moral course of action, is another question worthy of consideration but beyond the scope of this paper. It may well be that for almost all possible animal enhancements, there may be a more pressing obligation to do something that satisfies a competing human interest. It bears noting, however, that even within the realm of obligations to other humans alone, we have multiple and often competing obligations that must be balanced. Recognising that it is not feasible to fulfil all the obligations that might be incumbent upon us does not absolve us from the need to fulfil any of them; the mere presence of competing obligations does not render all obligations void.

The point of my argument here is not to establish that our obligations to enhance animals outweigh all other obligations such as those towards humans, but merely that such obligations exist and ought to be taken into consideration.

Returning then to our hypothetical mentally retarded human child and chimpanzee baby: do we have an obligation to enhance the chimpanzee? If not, why not?

First, is intelligence a benefit? The answer is probably, though not definitively, yes. Intelligence is generally considered to be a benefit to humans and studies show that it is correlated with other indicators of (insofar as these things can be measured) wellbeing (see for example Deary 2005; Gottfredson 2004). It would seem to serve an adaptive function in chimpanzees as in humans; and there are many ways in which we can imagine that a more intelligent chimpanzee may have a better life. At the least, there is more evidence in favour of it being a benefit than not. Of course, how a super-intelligent chimp might be treated would also affect the quality of its life; but this would be a reason to treat it in accordance with its intelligence, needs and interests, not to refrain from enhancing it in the first place!

If enhancement of intelligence is a benefit, therefore, we have a *prima facie* obligation to it. Of course, a great deal of work is required to demonstrate that cognitive enhancement is as much of a benefit to a chimpanzee as it is to a human; but that does not negate the existence of some obligation, even if it is not as great.

One might argue that because we are not the parents of the chimpanzee, there is no parental obligation. But most of us would probably feel that the obligation to the human child would hold even if we were not its parents – if we were doctors, or even just passers-by who happened to be in a position to do some good, we would have the obligation to do so. Why then should we feel there is an obligation to unrelated humans, but not to chimpanzees – who after all, are only slightly less unrelated to us than the average stranger, genetically speaking?

One might say that the human baby is at least our species, even if not a direct relative. Many philosophers in the area of animal ethics, though, have argued strongly and convincingly that speciesism is not a valid moral principle. Until we can find some morally plausible reason to enhance human animals but refrain from enhancing non-humans, we will have to give due consideration to their interests in enhancement and our correlative moral obligations to enhance.

What purpose should morality serve? I believe, as others have argued, that it should serve the survival not just of the individual or our biological species, but the survival of our moral species – creatures like us, not just in that they look like us or share a certain amount of DNA with us, but creatures who share the qualities we value in ourselves as humans, or at least elements of them.

Perhaps non-human animals are not quite yet capable of the kind of higher-order moral decision-making that most humans are. Let me propose, though, as both a closing point for this paper and a starting point for further discussion, that perhaps also it is our duty to help them become so.

Bibliography

Annas, G. (2001): Genism, Racism and the Prospect of Genetic Genocide. Presentation at UNESCO 21st Century talks (World Conference against Racism, Racial Discrimination, Xenophobia and Related Intolerance in Durban, South Africa, September 3rd.): http://www.geneticsandsociety.org/article.php?id=1994 (06.02.2012; 5.30 pm)

Benveniste, H. et al. (2007): Anatomical and functional phenotyping of mice models of Alzheimer's disease by MR microscopy. In: Annals of the New York Academy of Sciences 1097, 12-29.

Chan, S./Harris, J. (2007): In support of human enhancement. In: Studies in Ethics Law and Technology 1, Issue 1, Article 10.

Collins, J.J./ Evason, K./ Kornfeld, K. (2006): Pharmacology of delayed aging and extended lifespan of Caenorhabditis elegans. In: Experimental Gerontology 41, 1032-1039.

Deary, I.J./ Batty, D./ Gottfredson, L.S. (2005): Human hierarchies, health, and IQ. In: Science 309, 703.

Galas, D.J./ McCormack, S.J. (2003): An historical perspective on genomic technologies. In: Current Issues in Molecular Biology 5, 123-127.

Greely, H.T. et al. (2007): Thinking about the human neuron mouse. In: The American Journal of Bioethics 7, Issue 5, 27-40.

Gottfredson, L.S. (2004): Intelligence: Is It the Epidemiologists' Elusive "Fundamental Cause" of Social Class Inequalities in Health? In: Journal of Personality and Social Psychology 86, Issue 1, 174-99.

Harris, J. (2007): Enhancing Evolution. Princeton University Press, Princeton, NJ, US.
Knight, J. (2003): GloFish casts light on murky policing of transgenic animals. In: Nature 426, 372.
Lee, S.J. (2007): Quadrupling muscle mass in mice by targeting TGF-ss signaling pathways. In: PLoS ONE 2, e789: http://www.plosone.org/article/info:doi/10.1371/journal.pone.0000789 (06.02.2012, 5.30 pm).
Lee, S.J./ McPherron, A.C. (2001): Regulation of myostatin activity and muscle growth. In: Proceedings of the National Academy of Sciences USA 98, 9306-9311.
Lithgow, G.J./ Andersen, J.K. (2000): The real Dorian Gray mouse. In: Bioessays 22, Issue 5, 410-413.
Moravec, H. (1989): Human culture: a genetic takeover underway. In: Langton, C. G. (ed.): Artificial Life. The Proceedings of an Interdisciplinary Workshop on the Synthesis and Simulation of Living Systems Held September. Addison-Wesley, Redwood City, CA, Vol. VI, 167-199.
Muotri, A.R. et al. (2005): Development of functional human embryonic stem cell-derived neurons in mouse brain. In: Proceedings of the National Academy of Sciences USA 102, 18644-18648.
Ourednik, V. et al. (2001): Segregation of human neural stem cells in the developing primate forebrain. In: Science 293, 1820-1824.
Rose, M.R. (1989): Genetics of increased lifespan in Drosophila. In: Bioessays 1, 132-135.
Sandel, M.J. (2004): The Case Against Perfection. What's wrong with designer children, bionic athletes, and genetic engineering. In: Atlantic monthly 292, 50-54, 56-60, 62.
Sandel, M.J. (2007): The Case Against Perfection. Ethics in the Age of Genetic Engineering. The Belknap Press of Harvard University Press, Cambridge, MA.
Shultz, L.D./ Ishikawa, F./ Greiner, D.L. (2007): Humanized mice in translational biomedical research. In: Nature Reviews Immunology 7, 118-130.
Tamaki, S. et al. (2002): Engraftment of sorted/expanded human central nervous system stem cells from fetal brain. In: Journal of Neuroscience Research 69, 976-986.
UNESCO (1997): Universal Declaration on the Human Genome and Human Rights. UNESCO, Paris, France.

Ethical Assessment of Human Genetic Enhancement[1]

Nikolaus Knoepffler

The Meaning and Limits of the Topic

How is it possible to conceptualize the ethics relevant to the enhancement of human beings by means of genetic technology – genetic enhancement? The technical term "enhancement" eludes a precise definition, even though the English word enhancement is merely a lexical synonym for "improvement", which inherently means improvement. This raises the essential question regarding what constitutes a real improvement for human beings. Though one could dwell on the fundamental questions regarding whether it is even possible to improve a human person, and what would that improvement actually be, these questions will not be addressed in this paper. Rather, I begin with the proposition that it is possible to improve upon the human condition. An example from outside the field of genetics helps show what is meant here. The establishment of schools can be understood to be an attempt to promote and thereby to improve children not only in their cognitive, but also in their social capacities. Those school children who acquire new languages have not only mastered a specific cognitive skill, they have also broadened their potential for social engagement and professional accomplishment. This is not to deny the possibility that the acquisition of certain kinds of knowledge or cognitive skills can be of dubious advantage.

The topic can be further narrowed by concentrating upon what constitutes a genetic improvement of human beings. For example, how is it possible to construe the healing of a person from an illness with the use of gene-therapy as an improvement of the condition of the person? Should this fall under the category of enhancement? Or is this only a re-attainment of the person's "normal healthy basic condition"? This also raises a very important question: How are we to understand what constitutes a "normal healthy basic condition"? Is a medical intervention that stimulates the growth of an otherwise very small person a therapy or a dangerous consequence of social prejudices, as the Society of Little People of America claim (See Knoepffler 1999). This raises many questions that are in

---

[1] This article at some extent was previously published in Knoepffler (2009) and Knoepffler (2011).

themselves important, but they cannot be pursued here because the grey zone between the terms therapy and modification opens up an entirely new field of discourse similar to that discourse regarding the difference between modification and improvement-enhancement. Rather, for the purposes of this paper, I concentrate upon specific cases where the improvement from medical management constitutes a clear improvement. This follows from the above mentioned presupposition that it is possible to make the distinction between modification and improvement. For example, the utilization of genetic techniques to induce the production of human insulin in bacteria is clearly within the realm of therapy, and not just modification. So too is the use of a somatic gene-therapy for persons dealing with the effects of a very serious immune disease called SCID (severe combined immunodeficiency). Less clear are medical interventions which utilize genetically engineered products for increasing vision or for acquiring the range of visible perceptions which cats possess, such as seeing with very low levels of light. This last example clearly falls into the category of enhancement rather than treatment and is a bit of science fiction. However, there are a range of realistic opportunities for utilizing genetic engineering for the purpose of improving human physical performance. To eliminate the grey zones, further considerations will not deal with cases of therapeutic applications of genetic engineering in either a narrow sense or a wider sense. The use of genetic engineering for disease prevention and the elimination of risk factors is not a thematic concern here. Also excluded is the discussion on genetic processes such as pre-implantation diagnosis with the aim of achieving children with distinctly preferred characteristics, such as eye color, gender, resistance to disease, etc... The process of genetic selection, and not modification, falls outside of the concern of this study.

Conversely, what will play a crucial role is maintaining the distinction between genetic processes that are indirect or direct. Indirect processes describe the use of genetic engineering to develop materials and processes for treating human subjects. Direct is the use of technology to actually modify the genetic composition of a human being, and thereby achieve a medical outcome. This may occur on a somatic level, meaning that the genetic modification is limited to the person under treatment. Or it may occur on a germ-line level whereby the genetic modification is potentially transferrable to the treated person's future children. Another distinction can be drawn according to whether the genetic modifications (somatic or germ-line) are reversible. And a further distinction can be made on the basis of the form of genetic enhancement, whether it involves the advancement of physical capacities, the so-called physiological enhancement, or cognitive and social capacities.

In light of the above considerations, I will use the following definition for genetic enhancement:

Ethical Assessment of Human Genetic Enhancement 69

The concept "genetic enhancement" includes every form of physical, cognitive and social improvement for human beings with the use of genetic-technical methods for purposes that are either therapeutic or simply elective.

Thus the present paper will put together and then utilize an ethical framework to appreciate and respect the great complexity of the issue of genetic enhancement, but also to make the distinctions within the complexity clear. The following ethical structure aims to provide an understanding necessary to make sound ethical judgments. As a matter of simplification, I assume contra factually that the following technologies are well established and safe, and that no embryo research is needed. Without this assumption, the reflection upon every genetic procedure would be bogged down with the question of associated risk. This is a relevant concern in the concrete, but it is purely speculative and functions as a distraction from the question here regarding the ethics of the procedure itself. Procedures that carry less risk are obviously to be preferred for ethical and practical reasons. Yet the relevant ethical question here is whether it is ethically permissible, or even obligatory, that we undergo procedures with the goal of genetic improvement.

The Ethical Framework

Judgments in difficult bioethical cases are essentially dependent upon the ethical framework being utilized. Thus it is important to be quite clear from the beginning that the ethical framework being proposed here is built upon the principle of human dignity which has been established in international charters (UDHR) and national laws and which has achieved a distinct level of international consensus (see Knoepffler 2004, 7f). This principle was instituted in response to the cruel and inhuman beliefs of National Socialism, which rationalized the sacrifice of individual rights and personal dignity as necessary for the interests of the majority, and they instituted racist policies that denied the principle of human equality. The principle of human dignity stands as both a rejection of human debasement and a universal recognition of fundamental human worth. Every human being has value simple on the basis of being human, and upon that basis shares a fundamental equality with all other humans, as well fundamental human rights. This ethical framework is not essentially based upon any single philosophical or theological foundation, even when that is possible (see Knoepffler 2004, 33-49). The framework and its deep meaning is an historical achievement tied to the tragic experience of human degradation. This contingent context of the principle's creation does not imply that the principle itself is contingent. The context merely shows the background which lead to the formulation of an internationally recognized ethical principle.

Basic human rights are fundamentally bound to the principle of human dignity, as mentioned above. Thus, the rights to life and to bodily integrity are of central importance, as is the right to self determination. These two aspects of human rights, the physical and the volitional, frame the tension that exists in genetic enhancement debates: Do we have a right or at least permission to attempt to improve the basic core of who we are as human beings to promote our physical, cognitive or social well-being?[2]

In addition to this "objective" claim to life and bodily integrity, the principle of human dignity safeguards what might be considered a more "subjective" claim to freedom that is associated with the right to self determination. It is subjective in the sense that the person has the freedom to choose his own good, at least to the degree that these choices to not infringe upon the self determination of other people. The right to self determination prioritizes the freedom of individuals over and against the structures of potentially over-reaching, paternalistic or even dictatorial governing structures. Such protected freedoms can allow individuals to make bad or even tragic choices, but in modern liberal systems of law, the restriction of freedom is judged to be the greater evil. Subjective rights protect the individual in a sphere of freedom, within which the individual may act according to her or his own desires, provided that those actions do not infringe upon the rights and freedoms of others. Ideally, we should not act in ways that are detrimental to our well being. But according to the right to life and bodily integrity, we do not have an ethical duty to protect our lives and bodies to the greatest possible degree according to some "objective" criteria – even if that might be judged to be a meaningful and desirable thing. One need only look to the bodily risks involved in playing most sports to see this point. The positive aspects of engaging in sports far outweigh the risks involved, and in any case, the essential judging of those relative risks and benefits are generally left to individuals in modern liberal legal systems.

The present ethical framework conscientiously distinguishes itself from a Utilitarian approach that attempts to maximize the greatest possible good for the greatest number of people, or even sentient beings. This is not to deny certain teleological aspects of the system, but it is much more dependent upon a respect for human dignity and basic human rights, with an emphasis upon the right to

---

2 Cf. the World Health Organisation's statement that "the attainment of the highest standard of health (defined as complete physical, mental and social well-being) is one of the fundamental rights of every human being without distinction of race, religion, political belief, economic or social condition" (WHO 1946). Even though the WHO did not think of genetic enhancement, enhancement seems to fit very well with this statement.

life and self determination of all humans. These basic protections are necessary to realize the best possible life. The concept "best possible life" is not a standard that lends itself to objectification. Rather it is an open-ended goal that demands the respect of basic rights for all such that all may exercise their freedom to achieve what they judge to be their own subjective goods. There is an entire social aspect that is also relevant to this ethical framework for genetic engineering, but the social element builds upon a clear and unwavering respect for basic human rights and dignity.

Structuring Genetic Enhancement and Some Ethical Questions

Genetic enhancement is very often structured the following way:

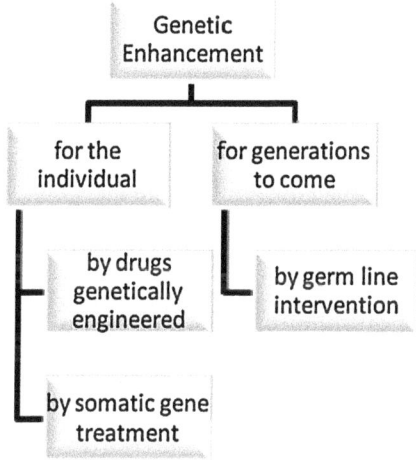

Based upon the preliminary ethical groundwork outlined above, the approach based upon the principle of human dignity leads to a conclusion that an individual may ethically receive a genetic substitution therapy, assuming that there are no unreasonable risks associated with the treatment. The indirect method of genetic substitution therapy is not qualitatively very different from a therapy relying upon medical processes that do not rely upon genetic engineering. For example, if the goal is to improve physical or cognitive abilities by means of drinking coffee specially brewed with a high level of caffeine, the ethics of drinking that coffee for that aim is independent of whether that coffee was brewed using "normal" or genetically modified beans.[3] Given that coffee brewing and drinking are socially acceptable practices, even in its concentrated espresso form, no ethical principle that is violated by the practice, given that it is not done in a context such as professional sports where it may be judged to be a form of doping agent.

A somatic interference in the human genome raises the ethical stakes considerably. If by changing the human genome one achieves an improvement, then the centrally relevant question involves the individuals who are affected by the action; does the improvement affect only the single individual receiving treatment, or is there a potential impact upon future children. Granting that the specific procedure was safe, the fundamental question remains regarding whether we are at liberty to make changes not only in our own germ line, but in our genome itself. Substitution therapy can be considered a rather harmless extension of our human potentialities outside the sphere of sports (such that we are not using the therapy as a means of cheating). Making changes in the genetic sequence itself, however, raises the stakes, as Habermas notes: "There is a proportion between the recklessness of modifying the human genome and the loss of respect incurred by the clinical establishment which attempts to utilize bio-technology for medical treatment. What is damaged is the intuitive differentiation between human development and fabrication, between the subjective and the objective – it is a matter that deals with human self-respect regarding our bodily existence. ... Intervention in the genetic processes of human beings suggests a dominance of nature in an action of self-empowerment, and it is one that not only changes our ethical self-understanding, it affects the very basic conditions for autonomous living and our universalistic understanding of morality." (Habermas 2002, 85) This is especially relevant when the modification of the parents' germ line leads to a kind of programming of their children. On this point, Habermas quotes Jonas: "But what is this power – and over whom and what exactly? Apparently it is a power of the "present" over the "future" which is defenseless against the decisions made prior to their coming into existence." (Jonas 1985, 168, quoted in Habermas 2002, 85). Habermas formulates this thought for himself in more careful terms: "The parents have made their own decisions according to their own lights and without consulting their children. Yet this is a matter that is central to the development of the person, and it is an egocentric intrusion into what should be considered a communicative act, and one that has potentially existential consequence" (ibid 90). The potential consequences include the limitations of the children's autonomy through the expectations of the parents.

---

3   There is an assumption that the ethics regarding the genetic modification of the coffee been, a separate issue, has been resolved.

A contradictory view is offered by Gerhardt (2004, 285) who argues that "A genetically manipulated man has, in fact, has his own insight, his own will and an ability to freely pursue his own desires. He can do all that he wishes – that he can grasp according his to own nature, even if his disposition is unique. Yet nature does not make a person unfree; rather the opposite is true. It is precisely nature's predictable order and laws that provide the necessary conditions of all freedom. Freedom, however, is unique to every person, and it depends upon nothing else so much as the recognition of person's own awareness of his situation in the world and then acting according to what he determines to be right and good." This point calls a further consideration from Habermas into question: "The reference to this difference between nature and culture, between unalterable beginnings and planned artificial processes makes it possible for actors to claim a certain performative self-understanding, without which he could not understand himself as the initiator of his own actions and claims" (Habermas 2002, 103).

It is very doubtful whether one can draw so clear a boundary between nature and culture, or between development and fabrication as Habermas tries to draw (see Köchy 2006, Reyer 2006). It is also at least theoretically possible to reverse modifications to the germ line. Though, for some distinct forms of socialization, modifications are inherently difficult to reverse. For example, the native language of a population is an inherently stable social practice. In any case, the reality of genetic modification is one that is open to differentiation, and even Habermas is explicitly open to an exception for therapeutic aims (2002, 119), though his argumentative style does not lend itself to easy understanding, the exception for therapy still implies the modification of the genetic code). So even he would admit that it is permissible to modify the genes of a human gamete with a trisomy at the eight-cell level (an abnormality consisting in an extra set of genes). The disposal of the abnormal gene on the eighteenth chromosome (trisomy 18) at this cell level would be necessary for the survival of the fetus after birth.[4] For humans with trisomy 21 (Down syndrome), if future research makes it possible, the timely modification of the gene could result in greater freedom.

The groundwork having been laid, we may now proceed to a deeper understanding of the ethics of genetic enhancement. According to the provisions of the ethical framework, if the enhancement results in a clear and defensible improvement, it can withstand the attack of ethical criticism that it leads to a restriction or removal of autonomy, as Jonas and also Habermas have suggested. A clear

---

4 This position can be taken by theologians, too: „Nach Auffassung vieler [katholischer] Ethiker wäre grundsätzlich ethisch nichts dagegen einzuwenden, eine genetisch bedingte Anomalie in einem menschlichen Embryo auf gentechnologischem Weg zu korrigieren" (Erwachsenenkatechismus der Deutschen Bischofskonferenz 1995, 301).

and defensible improvement, however, must consist in a true expansion of the person's freedom and potential to act. Such an expansion corresponds precisely with basic human rights, which are the essential pillars and symbols of freedom.

It is evidently conceivable that people will complain about the expansion of their potential possible actions, and obviously nobody can be certain that this expansion will result in an "objective" improvement in the actual lives of human beings which are experienced subjectively. But this argument does not carry much weight. There will probably always be people who are unhappy with their lives, who do not choose to exercise much of the vast freedoms they have, or who even wish that they had never had been born.[5] Rather, if it can be demonstrated that a modification of the human genome constitutes a very probable improvement of human life, then there is no fundamental reason inherent in the action of genetic modification which serves to prohibit it.

Thus the real task in dealing with this question is to get beyond the generalities of such questions as whether every intervention of genetic engineering for the purpose of enhancement should be forbidden, or even every modification of the genome for enhancement should be forbidden. Rather, the real task is to differentiate the distinct levels of enhancement so that the ethical decisions can be made. This differentiation is essentially ordered according to the degree to which a genetic modification expands or contracts a person's freedom and ability to act, thus sustaining and strengthening the person's fundamental right to physical, cognitive and social well being. Here, Habermas' emphasis upon personal autonomy can play an important role.

An Improved Structural Model of Genetic Enhancement

In sum, the following model proceeds from the assumption of the principle of human dignity, from the sense of the human person as a having "subject status" prohibiting the use of persons as mere means, from a belief that the all persons share a basic equality with all others, and from a careful consideration of the person's right to self determination. Given all this, it is essential to take into account whether the decision to undergo a procedure involving genetic engineering was made:

by the person themselves
or by the parents
or by a specific group in society
or by authorities in local or more general government.

---

5   Cf. the critique of Harris (2007, 137-142) in respect to Habermas.

Ethical Assessment of Human Genetic Enhancement 75

There are good reasons why these decisions should not be entrusted either to groups in the society or to governmental authorities. Even though there is a role for the state to play in the raising and education of children through regulation such as school requirements, they ought to have no regulative role concerning genetic therapy, because it would constitute too great an interference into individuals' right to self-determination. Although the state has the obligation to provide protections, it may not prescribe what we ought to drink or eat and it has even less of a right to prescribe how individuals ought to be genetically modified. The fundamental subject status of every individual sets clear boundaries limiting the government's role regarding genetic enhancement to at most an advisory one. Its recommendations, however, should not extend so far that they might constitute incentives that indirectly and perhaps patronizingly influence citizens. Lying behind this caution is the conviction that no state or social organization has the right to exercise direct or even indirect compulsion to form citizens or members of an organization according to predefined criteria. Such a scenario would overturn the necessary priority of the individual in the state. The state exists to serve its people, and the subject status of all members of the state protects them in their basic biological conditions, as is necessary according to the principle of human dignity.

Similarly, the decisions of individuals should not be unduly influenced by social groups regarding these matters. Social groups, like the state, have their proper and necessary roles to play. This clearly does not exclude the important advisory role of religious communities, for example. People freely associate themselves as members of a religious community and they do not necessarily compromise their autonomy when acting according to a religious "authority". However, the following principle should determine practice: The right to self-determination should function in a manner analogous to the self-determination that rules in the field of therapeutic medicine, namely, that the patient has the definitive last word. Further, it means that people or groups may not make decisions for an individual if they do not belong to the individual's freely chosen decision-influencing circle.

A much more difficult question is posed when parents exercise their responsibility to make decisions for the children. Therefore, I would like to offer a preliminary and clarifying differentiation between:

a decision for the parents' own benefit, but one that carries implications (positive or negative) for their children, and

a decision of the parents for the benefit of their children and possibly for further generations.

What does "possibly" mean? The "decision of individuals for themselves" entails that the genetic enhancement is intended to affect only the treated individual. A so-called parental decision is one which means that the modification of the genome will very probably involve implications for further generations. Such a decision raises the level of ethical implications and justifies more restrictive measures to protect all those potentially affected by the decision, including those not yet been born. One must consider all the people who are directly affected by the genetic enhancement: Does it only affect the person being directly treated, as is the case in a genetic substitution therapy or a somatic gene treatment? Or does the intervention of genetic engineering affect further generations of the person under treatment, which is potentially the case when there are modifications of the germ line.

Next, we must consider the objective or purpose which motivates the intervention of genetic engineering: Is it intended to improve physiological or bodily characteristics? Should it increase cognitive capacities? Or should it improve the person's behavior? This leads us to the following structure:

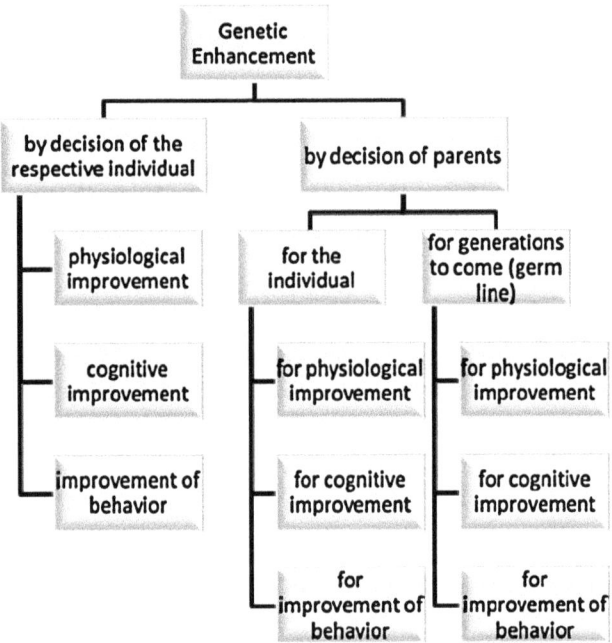

Ethical Cases for Consideration

The following three case studies, as mentioned above, will assume for the purpose of simplification and clarity that the applied procedures would achieve a real improvements. They are fictional characterizations intended to illuminate the potential ethical distinctions to be made among potential applications of genetic engineering for the purpose of enhancement. They are intentionally peculiar, and in no way reflect current medical practices or research for future research.

First Case Study: Improving the Human Eye

Assume that it would be possible, without side effects, to modify our genetic constitution in such a way that we could achieve a low-light vision like cats as well as an eagle's sharpness of vision. Assume further that this modification would entail no harm to our normal human vision. It would not cause any damage to our capacity to read and write, just as before the modification. Given these assumptions and conditions, would there be any ground to object to such a modification of the human genetic constitution?

Let us begin by considering this case in terms of a genetic modification that only affects the treated individual, i.e., the improvement of sight is achieved by means of a somatic genetic enhancement. An objection might be raised that this violates the distinction between development and fabrication. Indeed, Habermas had associated this objection with the question of self-determination. Yet, the modification of our ability to see could possibly expand our freedom and ability to act, and it is not obvious how it would limit our self-determination. The electron microscope and the telescope manifestly improved the capabilities of humans to see things very small and very far away (see Harris 2007, 19f). The employment of special glasses has become commonplace to protect our eyes from the Sun's brightness and UV rays and thereby expanded our capacities for action. It seems that the modification of a person's own capacity for sight would achieve similar capacity expansion. Questions might be raised at this point regarding the effects of such an improvement to sight. Perhaps we would see things that we do not want to see. In such a case, we as individuals could choose, subjectively, whether this modification constitutes a true improvement. If we decide that it does not, then we retain our freedom to refuse such a modification.

A common objection is that such modifications are "unnatural" in the sense that they contravene what is considered the given order of nature, whether that is based upon a religious belief in "creation" or a non-theistic worldview. Yet it is difficult to establish a clear standard of valuation from what is given in nature. How "natural" is the use of sunglasses to protect against UV rays? How "natural" is our industrially and culturally complex contemporary life? Even given the possibility that a rationally defensible ethical measure could be determined, to what degree is it possible that such a system could be established as mandatory for people not sharing its basic philosophical assumptions? Thus, according to the criteria presented here, such a genetic enhancement would potentially expand human freedom and potential for action. Therefore, such a modification could not be legitimately restricted, though it would obviously remain a voluntary option.[6]

A further objection might be raised on the level of justice or fairness: If such a modification were available only to a select group of people by high cost or some other reason, would it create a new elite and thus social unbalance? Is one person affected adversely when other people suddenly possess an elevated capacity such as powerful vision? What does it mean concretely if a genetically modified person is capable of actions for which people had previously required mechanical measures? Certainly that person possesses an advantage in competition for employment over and against a non-modified person. A resolution to this problem could be achieved in way analogous to the measures society takes to make education opportunities open to all in society (cf. Buchanan at al. 2001; Harris 2007). This resolution is admittedly only applicable to the case on this first level, but a system could be built to neutralize the unbalances that exist based upon economic and social differences. The existing education system could serve as a model, where state-funded primary and advanced schools provide universal access to education, within certain limits.

And what happens to those people who refrain from genetic modification? They certainly have such a right according to the principle of self-determination, but they face a situation analogous to those who do not pursue a publically available and sponsored educational opportunity that opens a corresponding career field. John Rawls, considered by many to be the most important philosopher of justice in the twentieth century argued that it is in everybody's interest that each individual receive diverse but significant natural gifts. "It is in the interest of each to have greater natural assets. This enables him to pursue a preferred plan of life. In the original position, then, each party wants to assure for his descendents the best genetic endowments" (2005 [1971], 108). "have the most The diversity and significance of individual gifts aid not only the receiver of the gifts pursue his own goals, which include passing benefits on to children, but there is also a knock-on effect to others in society. Rawls is admittedly not speaking about genetic enhancement here, but his theory of justice does not exclude the possibility of, and the general desire for, a diversity of abilities as comprising a social good.

---

6   Theologically one can defend this position if one understands the act of creation not as a single defined act, but as a continuous process, in which the human being contributes as concreator: „Das Eingreifen des Menschen in die Natur und Umwelt ist Auftrag Gottes; es ist nicht nur sittlich vertretbar, sondern es ist auch sittlich verpflichtend" (Deutsche Bischofskonferenz 1995, 299).

The issue becomes more complex when dealing with the improvement of sight with genetic modification, and the process involves a modification of the genome itself. The stakes are raised because the process affects a greater number of people and the potential for reversing the process is reduced. Yet, people make numerous decisions which affect their children, such as choosing a place to live, choosing early schools, passing on religious beliefs, etc. There are significant consequences to many decisions that parents have to make, and we take it for granted that parents have a right to assume this responsibility. The capability for reversing the modification in later generations makes the case a little easier. Yet even if this is not the case, there should not be a presumption that children are harmed simply because they have not given consent to a procedure that affects them. And if the modification has already been shown to be safe and one that advances the potential for self-determination, as shown above, then it could also be understood as desirable.

The attribution of desirable, as opposed to obligatory, differentiates the position here from the utilitarian approach of those like John Harris (2007, 19ff). The designation of desirable demonstrates this system's foundation in human dignity and self-determination, as the individual is free to choose that which is determined to be a good, but is not bound by duty to choose that which is reasoned to be the best good. A relevant consideration might include the anticipated wishes of subsequent generations. It is reasonable that one may refrain from undergoing a genomic modification, especially when the future option for modification remains open.[7] What might be reason for refraining from a risk-free genetic enhancement? One reason might be existential, namely, a position that the purpose of human existence is not to attain physiological perfection, but rather spiritual fulfillment. Fulfillment might, furthermore, include an acceptance of a limited human existence with aspects of both dark and light, joy and pain, power and weakness. With a foundation upon the principle of human dignity, the moral framework proposed here must respect such existential convictions and the freedom of individuals to act according those convictions.

The Case Study of Improving Memory Performance

Genetic engineering is potentially effective in modifying and improving not only physiological, but also cognitive capacities such as memory. In contrast to the first case study, which was somewhat fantastical, there are already animal expe-

---

[7] In contrast gene therapeutic modifications are a duty towards subsequent generations if they prevent future children from getting life threatening diseases.

riments in progress testing short-term memory. Similar to the first case study, however, it seems that a genetic modification to improve short-term memory (presuming it has no negative side-effects[8]) must be presumed to be an improvement worthy of support. This is the case whether it involves a somatic or a genomic procedure because no convincing argument can be found to prohibit it. Also, it should be presumed that future generations would support the expansion of their short-term memory capacities.

Köchy (2006, 81) quite wisely argues regarding the present options, "the more complex the phenotypic features, whether they refer to intelligence, musical talent, bravery, creativity ... etc., the greater the proportion of other genetic and environmental factors in the realization of that trait. For this reason, the targeted design with regards to humans suffers from effective and technical restrictions." This reality is not apparent in the animal experiments that tend to concentrate on the correspondence of a single gene or gene group, to a single trait. Köchy's reservations must be taken seriously, but for the purpose of this argument, would it be permissible or even praiseworthy to undergo a process if it was indeed possible to improve short-term memory without such effective and technical restrictions?

One could argue for such a ban in the following manner. Although an improvement of the short-term memory is directly advantageous, simply because it is much better to remember things, such a modification would have broader consequences than those that occur on a simply physiological level. A change of memory would affect human personality. This wide-reaching and profound intervention is one that should not be available for parents. This restriction might raise the following objection: Why should a genetic modification for the cognitive improvement of short-term memory possess negative consequences? Could it not also be argued that an individual undergoing such a process could also gain significant benefits? The case studies have proceeded to this point under the counter-factual assumption that the procedures would be safe. Is it enough to assume that a potential personality change is enough to prohibit this potential practice? If this practice, as with all medical practices, proves to have negative consequences, then those must be factored into its ethical evaluation. But if it is granted that a genetic enhancement to improve short-term memory in fact poses an improvement without significant negative consequences, and if such an improvement entails an expansion of a person's freedom and self-determination, then is seems that a prohibition is not well grounded. This conclusion is thus consistent with the principle of human dignity.

---

8   At moment, physicians would prescribe Ritalin and would not recommend risky genetic memory enhancement.

The Case Study regarding Changing the Behavior of a Pedophile

This case is much more difficult than the former cases as it involves the use of genetic engineering to modify behavior. This is the case because it poses the question whether a behavioral change that is sufficiently complex to resist therapeutic treatment could actually be improved. Winnacker (2002, 62f) and others have come to the following conclusion: this application of technology to human beings, this genetic engineering for the production and refinement of human beings, would be an extreme undertaking. This is because in our pluralistic world, we are supposed to following a clear rule regarding when we may make an intervention and when we may not. But in this case, how are we supposed to find a secure position that will remain valid for coming generations? It is in no way clear, which form of intelligence or social behavior will be preferred in the future. Such an intervention does not lend itself to general and scientifically defensible foundations and criteria. A science that would adopt such goals risks making the scientist into a superhuman designer of humans."

This judgment regarding the modification of human behavior with genetic engineering does not take deal with the elimination of very negative behaviors such as pedophilia. There can be little doubt that a genetic enhancement that clearly and verifiably prevents pedophilia behavior would be an unqualified good. It must be granted that it is by no means clear to what degree pedophilia can be traced to genetics, and that behaviors are very unlikely to have a mono-causal relationship with single or even multiple gene sequences. But the question regards not the natural sciences, but philosophy and ethics. If it could be granted that a person who possesses a compulsive desire to abuse children, i.e., a pedophile, could be undergo a treatment involving a genetic modification of their genome, would that be ethical? Keep in mind that such a modification would affect not only the single treated person, but also future generations such that they would not exhibit such tendencies. If such a treatment were to be available, would the protection of children strongly incline society to allow it or even promote it?

It is against the background of the principle of human dignity, Sloterdijk (1999, 44f) agrees: "When an intellectual power has achieved a level of authority, humans should not cower and balk as they have done in earlier periods of powerlessness, when they have surrendered their responsibilities to such higher powers as God or chance or simply other humans. In order to prevent such a simple refusal or resignation in the future, we should pick up the ball ourselves and formulate a codex of anthropological technology. Such a codex would retroactively change the meaning of classical humanism. Humanitas, as it was so often written, finds fulfillment not only in the friendship of humans, one with another,

but it also implicitly – and with increasing explicitness – includes the reality that humans gather to themselves the greatest power."
The task of dealing specifically with the possible ethical applications of genetic engineering is an admittedly difficult and protracted task. It is much harder that declaring a simple "no" or a simple "laissez faire." The construction of such a codex that differentiated the permissible and impermissible practices would require a painstaking and careful work for it to be accepted by society and effective scientifically. For such a codex, the measure would have to remain the principle of human dignity with the human rights that are bound to it. Crucially important is the criteria whether a genetic intervention to achieve a physiological, cognitive or behavioral modification could offer an expansion of freedom to act, i.e., to exercise one's self-determination. This is the criterion which should signal a true genetic improvement, i.e., an enhancement. As such, it deserves not to be forbidden, but should be available to individuals for their own good and for the good of their children.

Bibliography

Agar, N. (2004): Liberal Eugenics. In Defence of Human Enhancement. Blackwell, Oxford.
Birnbacher, D. (2006): Natürlichkeit. de Gruyter, Berlin
Buchanan, A. et.al. (2001): From Chance to Choice. Genetics and Justice. Cambridge University Press, Cambridge.
Busch, R. et.al. (2002): Grüne Gentechnik. Ein Bewertungsmodell. Herbert Utz Verlag, Munich.
Deutsche Bischofskonferenz (Ed.) (1995): Katholischer Erwachsenenkatechismus II. Leben aus dem Glauben. Herder, Freiburg im Breisgau.
Gerhardt, V. (2004): Geworden oder gemacht? Jürgen Habermas und die Gentechnologie. In: Kettner, M. (ed.): Biomedizin und Menschenwürde. Suhrkamp Verlag, Frankfurt am Main, 272-290.
Habermas, J. (2002): Die Zukunft der menschlichen Natur. Auf dem Weg zu einer liberalen Eugenik? 4th edition. Suhrkamp Verlag, Frankfurt am Main.
Harris, J. (2007): Enhancing Evolution. The Ethical Case for Making Better People. Princeton University Press, Princeton.
Jonas, H. (1985): Lasst uns den Menschen klonieren. Von der Eugenik zur Gentechnologie. In: ibid.: Technik, Medizin und Eugenik. Praxis des Prinzips Verantwortung. Suhrkamp Verlag, Frankfurt am Main, 162-203.
Kitcher, P. (2004): Creating Perfect People. In: Burley, J./Harris, J. (eds.): A Companion to Genetics. Blackwell, Oxford, 229-242.

Ethical Assessment of Human Genetic Enhancement 83

Knoepffler, N. (1999a): Ethical Evaluation of Different Levels of Gene Therapy. In: Nordgren, A. (ed.): Gene Therapy and Ethics (Acta Universitatis Upsaliensis: Studies in Bioethics and Research Ethics). Uppsala Universitet, Uppsala, 151-158.
Knoepffler, N. (1999b): Forschung an menschlichen Embryonen. Was ist verantwortbar? Hirzel, Stuttgart.
Knoepffler, N. (2004): Menschenwürde in der Bioethik. Springer, Heidelberg.
Knoepffler, N. (2007): Der moralische Status des frühen menschlichen Embryos. In: Diedrich, K. /Hepp, H. /von Otto, S. (eds.): Reproduktionsmedizin in Klinik und Forschung. Der Status des Embryos. Deutsche Akademie der Deutschen Naturforscher Leopoldina e.V. Wissenschaftliche Verlagsgesellschaft, Stuttgart.
Knoepffler, N. (2009): Ein Strukturmodell genetischen Enhancements. In: Knoepffler, N./Savulescu, J. (eds.): Der neue Mensch. Enhancement und Genetik. Karl Alber Verlag, Freiburg im Breisgau, 277-196.
Knoepffler, N. (2011): A De-escalation Model of Human Genetic Enhancement. In: Schleidgen, S. et.al. (eds.): Human Nature and Self Design. Mentis, Paderborn, 137-153.
Köchy, K. (2006): Gentechnische Manipulation und die Naturwüchsigkeit des Menschen. In: Sorgner, S.L./Birx, H.J./Knoepffler, N. (eds): Eugenik und die Zukunft. Karl Alber Verlag, Freiburg im Breisgau, 71-84.
Rawls, J. (2002 [1971]): Eine Theorie der Gerechtigkeit. 12th edition. Suhrkamp Verlag, Frankfurt am Main.
Reyer, J. (2006): Pädagogik in einer eugenisierten Gesellschaft. In: Sorgner, S.L./Birx, H.J./Knoepffler, N. (eds): Eugenik und die Zukunft. Karl Alber Verlag, Freiburg im Breisgau, 177-199.
Sorgner, S.L. /Birx, H.J. /Knoepffler, N. (eds.) (2006): Eugenik und die Zukunft. Karl Alber Verlag, Freiburg im Breisgau.
Winnacker, E.-L. et.al. (2002): Gentechnik: Eingriffe am Menschen. Ein Eskalationsmodell zur ethischen Bewertung (deutsch/englische Ausgabe). 4th edition. Herbert Utz Verlag, Munich.

Evolution, Education, and Genetic Enhancement

Stefan Lorenz Sorgner

We have reached a stage in history at which it seems possible to be an active part in the process of human evolution. We can use autoevolution to enhance the process of evolution. This type of enhancement can occur in a genetic or in a non-genetic way. In the first part of the article, I will distinguish four types of enhancement by means of which a posthuman can come into existence: Genetic enhancement; Non-genetic enhancement by means of education; Non-genetic enhancement by means of e.g. drugs; Non-genetic enhancement by means of the invention of Cyborg technologies (cyborg = cybernetic organism), e.g. by means of establishing mechanical or digital human-machine interfaces.

Various types of enhancement can bring about changes which can get inherited and which therefore have the potential to enhance evolution so that the posthuman can come about, as I will show. The decision in favour of the process of genetic and non-genetic enhancement can be made autonomously or heteronomously. As actions related to autonomous decisions are usually morally less problematic than heteronomous ones, I will take the latter into consideration in part two of the article, in particular genetic enhancement by selection and by modification whereby the decision has to be made by parents for their children.

I focus on genetic enhancement, because I regard it as the most likely option for bringing about the posthuman, and for having an effect on some processes of human evolution, so that natural selection receives further support from the process of human selection. By showing that there are structural analogies between both genetic enhancement by selection as well as genetic enhancement by modification, I wish to put forward some reasons for claiming that genetic enhancement does not have to be seen as morally problematic.

Enhancement and the Posthuman

The "posthuman" is a concept which comes up in various discourses both in the Anglo-American as well as in the Continental world, if one wishes to stick to this far from clear cut and blurred distinction. In the English speaking discourse, it is particularly prominent within the transhumanist movement. Within the literary and continental tradition, it comes up within posthumanism. Transhumanism is connected more closely with the enhancement debate which takes place in the English speaking world among analytical ethicists. Posthumanism is a movement which is more closely connected to the so called continental tradition of philosophy, and there is a close link between posthumanist and postmodern thinkers. However, transhumanism and posthumanism have in common that they both reject the categorically special status of human beings which has been connected with humanism.

Transhumanism affirms technological means in order to alter human beings which are for them "works in progress" to bring about the transhuman or the posthuman (Bostrom 2005, 1). The meaning of the concepts of the trans- and the posthuman differs significantly among transhumanist thinkers. However, transhumanism upholds the fully rounded personality as an ideal which is similar to one type of the Renaissance ideal. Hence they affirm a type of ethical humanism.

Posthumanism, on the other hand, is characterised by the dissolution of absolute moral standards, and a type of perspectivism, and aims for a new this-worldly anthropology. Posthumanists do not necessarily have any fundamental objections against technologically altering human beings. However, they do not uphold the absolute validity of the Renaissance ideal. As in transhumanism, there are concepts of the posthuman within the posthumanist discourse, too, e.g. the one from Katherine Hayles' in "How we became Posthuman" (1999) or the cyborg of Donna Haraway's "A Cyborg Manifesto" (1991, 1149-181). Sloterdijk is another philosopher who can be described as posthumanist, as he employs the concept posthumanism in a special way in his speech „Regeln für den Menschenpark" (2001, 302-337). A talk on the optimization of human beings which he gave at the 06th of December 2005 at the University of Tübingen represents a further posthumanist perspective, as here he is as critical concerning genetic enhancement methods as most of the other leading representatives within the traditional German discourse today which focuses on the concept of human dignity like Spaemann or Habermas.[1]

The posthuman in the transhumanist discourse can refer to a new step in evolution. The posthuman, in this case, would be a member of a new species. Other transhumanists uphold that the posthuman merely has qualities which go beyond the qualities of currently living human beings, but cannot be regarded as a member of a new species. Transhumanists are often discussing genetic enhancement,

---

1   See DVD from Peter Sloterdijk which is entitled „Optimierung des Menschen?" (2005, Quartino GmbH).

Evolution, Education, and Genetic Enhancement 87

Cyborg enhancement, and enhancement by means of drugs/medicine as methods for bringing about the posthuman.

Within the posthumanist discourse, the concept of the cyborg, the hybrid and the posthuman comes up, too. In their context, it represents a new anthropology. None of the leading posthumanists so far regards the realisation of a super-cyborg, an over-hybrid, or a post-posthuman as a goal worth aspiring for. Hence, the question concerning the future development of human beings in an evolutionary sense is not one of the central questions for posthumanists.

The various uses of the term "posthuman" are problematic, because they brought about many misunderstandings so far. I am affirmative of the posthumanist project of realising a new non-dualist understanding of human beings in the posthumanist sense. However, their usage of the term posthuman is definitely misleading, because it already gets associated widely in the sense that it refers to a further developed human being, as transhumanists use the concept. Hence, it might be advisable to choose a different term for referring to a non-dualist understanding of human beings, e.g. the metahuman. In this case the trans- and the posthuman can still be used in the sense FM 2030 used the concepts. The transhuman is a human being who still belongs to the human species but is on the way of turning into a posthuman. The posthuman, on the other hand, no longer belongs to the species of human beings. It is not yet possible to specify the description of the posthuman further, because it can refer to a mostly natural entity which can be realised by means of education, enhancement drugs or genetic enhancement. On the other hand, it cannot be excluded either that the posthuman comes about via the Cyborg and exists solely in the digital realm. This is the way I am employing the concepts meta-, trans- and posthumans. Hence, this line of development can be seen as a way of describing the various stepts away from the current human constitution: the metahuman.

I also think that from a posthumanist's perspective it ought to be an open question whether it is morally problematic to aim for the realisation of becoming such an enity, a posthuman, a member of a new species. I personally doubt that that it is a moral obligation to support the realisation of the posthuman, because such a claim might have implications of a-not-so-new-eugenics. However, I do not think either that it has to be morally problematic to have this goal as a personal aim. It can be a legitimate choice to support the process towards the posthuman. Yet, whoever chooses not to promote the posthuman does not act immoraly. This does not mean that I cannot understand that the posthuman can be an entity whose potential for leading a good life might be higher than that of the metahuman, e.g. of currently living human beings.

A further option for bringing about a posthuman, which has not been in the focus of transhumanists is by means of education. This option was put forward by

Nietzsche concerning our development towards the overbeing, the *Uebermensch*, as he upheld a version of Lamarckism. (Sorgner 2009, 2010, 2011)

Given the options mentioned, I can summarise that the posthuman can be brought about by means of human selection in the following ways:

We can enhance evolution by means of human selection, if we employ *genetic enhancement*. It might be possible eventually to alter genes such that even a new species can come into existence. It is possible that the decision for such an alteration will be made autonomously or heteronomously.

In addition, we can also bring about the posthuman by recourse to non genetic means of enhancement:

Nietzsche put forward education as a means to bring about the posthuman. Can education bring about changes which have an influence on the potential offspring of the person who gets educated? As inheritance depends upon genes, and genes do not get altered by means of education, one used to believe that education cannot be relevant for the process of evolution. Hence, Lamarckism, the heritability of acquired characteristics, has not been very fashionable for the same period of time. However, in recent decades doubts have been raised concerning this position which was based upon recent research on epigenetics. Together with Japlonka and Lamb, I can stress that "the study of epigenetics and epigenetic inheritance systems (EISs) is young and hard evidence is sparse, but there are some very telling indications that it may be very important". (Japlonka/Lamb 2005, 248).

Besides the genetic code, the epigenetic code, too, is supposed to be relevant for creating phenotypes, and it can get altered by means of environmental influences. The epigenetic inheritance systems belongs to three supragenetic inheritance systems which Japlonka and Lamb distinguish who also stress that "through the supragenetic inheritance systems, complex organisms can pass on some acquired characteristics. So Lamarckian evolution is certainly possible for them" (Japlonka/Lamb 2005, 107).[2]

Given recent work in this field it is likely that stress[3], education[4], drugs, medicine or diets can bring about epigenetic alterations which again can be responsi-

---

2 "Heritable variation – genetic, epigenetic, behavioural, and symbolic – is the consequence both of accidents and of instructive processes during the development." (Japlonka/Lamb 2005, 356)
   A striking case is that of the evolution of language:
   "Dor and Japlonka see the evolution of language as the outcome of the continuous interactions between the cultural and the genetic inheritance system." (Japlonka/Lamb 2005, 307)
3 "Waddington's experiments showed that when variation is revealed by an environmental stress, selection for an induced phenotype leads first to that phenotype being induced more frequently, and then to its production in the absence of the inducing agent." (Japlonka/Lamb 2005, 273)
4 Jonathan M. Levenson and J. David Sweatt show that epigenetic mechanisms probably have an important role in synaptic plasticity and memory formation (2005, 108-118).

ble for an alteration of cell structures (Japlonka/Lamb 2005, 121) and of the activation or silencing of genes (Japlonka/Lamb 2005, 117).[5] In some cases, the possibility cannot be excluded that such alterations can lead to an enhanced version of evolution. Japlonka and Lamb stress the following:

> "The point is that epigenetic variants exist, and are known to show typical Mendelian patterns of inheritance. They therefore need to be studied. If there is heredity in the epigenetic dimension, then there is evolution, too". (Japlonka/Lamb 2005, 359).

They also point out that "the transfer of epigenetic information from one generation to the next has been found, and that in theory it can lead to evolutionary change" (Japlonka/Lamb 2005, 153). Their reason for holding this position is partly that "new epigenetic marks might be induced in both somatic and germ-line cells" (Japlonka/Lamb 2005, 145).

A "mother's diet" can also bring about such alterations, according to Japlonka and Lamb (2005, 144) hence the same potential, as the ones stated before (genetic enhancement and education), also apply to the next method of bringing about a posthuman, i.e. non-genetic enhancement by means of drugs, medicine or diets. As it has become clear already, such measures can lead to an enhanced version of evolution, given recent research in the field of epigenetics.

Finally, there is the option of creating posthumans by creating human-machine interfaces or cyborgs. The flourishing research concerning bionics is concerned with this option. Cyborgs stand for interfaces between digital or mechanical machines and organic, natural beings, e.g. human beings/metahumans. It is still an open question, what their potential for human evolution is? It can be thought, and some futurists like Ray Kurzweil have played around with this idea, that one can download the content of ones mind to the hard disk of a pc or upload digital content to ones mind. If it was possible to realise these procedures, they might have the potential of bringing about a completely new concept of evolution which could be described as autopoietical digital evolution. However, this option is still far from being actually realisable which is the reason why I will not be concerned with it here – even though I am not excluding the possibility that such a procedure can be realised. Still, it seems to me that human machine interfaces in the near future do not yet play a role concerning human evolution but merely represent an enhanced metahuman that cannot pass his further developed capacities on to another generation. Genetic enhancement, on the other hand, seems far more promising for the bringing about of the posthuman.

---

5 "Belyaev's work with silver foxes suggested that there is a hidden genetic variation in natural populations . This variation was revealed during selection for tameness, possibly because stress-induced hormonal changes awakened dormant genes." (Japlonka/Lamb 2005, 272).

As we have seen so far various types of enhancement can bring about changes which can get passed on to another generation and which therefore have the potential to enhance evolution so that the posthuman can come into existence. The decision in favour of the process of enhancement can be taken either by oneself, autonomously, or by a person x for a person y, i.e. heteronomously. As actions related to autonomous decisions are usually morally less problematic than heteronomous ones, I will focus on the latter option in part two of the article, whereby my particular focus will lie on genetic enhancement. Thereby, one has to distinguish clearly between genetic enhancement by means of selection (after IVF, in vitro fertilisation, and PGD, prenatal genetic diagnosis) and genetic enhancement by means of alteration or modification. Both procedures can be named genetic enhancement. Firstly, I will concentrate on genetic enhancement by means of selection to provide some reasons for holding that it is structurally analogous to choosing a partner for procreative purposes, and that both procedures ought to be evaluated morally analogously, too. Secondly, I will deal with genetic enhancement by means of alteration or modification and classical education. As I also regard these procedures as structurally analogous, morally they ought to be evaluated according to the same standarts, too. By revealing structural analogies between traditional procedures and new ones which have come about due to the invention of emerging technologies, I provide an initial basis how one ought to deal with them morally. If my suggested structural analogies are plausible interpretation, then there is a reason for claiming that genetic enhancement procedures do not have to be seen as morally problematic.

Genetic Enhancement by means of Selection and Choosing a Partner for Procreative Purposes: A Structural Analogy[6]

In this passage, I will provide some initial reasons for holding that between the case of genetic enhancement by means of *selecting an already given genetic makeup* and the process of selecting a partner with whom one wishes to have offspring a structural analogy is given. I will clarify this point further in a more detailed analysis of this thesis.

By choosing a partner with whom one wishes to have offspring, one thereby implicitly also determines the genetic makeup of one's kids, as 50 per cent of their genes come from one's partner, and the other 50 per cent from oneself. By selecting a fertilized egg, one also determines 100 per cent of the genetic makeup by means of selection.

---

6   See Sorgner 2011, 21-25.

Evolution, Education, and Genetic Enhancement 91

One objection, which might be raised here, is that selecting a fertilized egg cell is a conscious procedure but normally one does not choose a partner according to their genetic makeup such that one has specific genes for one's child. However, it can be objected that our evolutionary heritage might be more effective during the selection procedure of a partner than we consciously wish to acknowledge. In addition, the qualities according to which we choose a fertilized egg after a PGD might not have been chosen as consciously as we wish to believe, but might be influenced more on the basis of our unconscious organic setup than we wish to acknowledge. It might even be the case, that the standards for choosing a partner and for choosing a fertilized egg might both be strongly influenced by our organic makeup and evolutionary heritage such that both are extremely similar.

A difference between these two selection procedures is surely that in the one case, one selects a specific entity, a fertilized egg, but in the other case a partner and therefore only a certain range of genetic possibilities. However, given the latest epigenetic research, we know that genes can get switched on and off, which makes an enormous difference on the phenomenological level. Hence, it is also the case that by choosing a fertilized egg, we only choose a certain range of phenomenological possibilities of the later adult, as is the case by choosing a partner for procreative purposes.

The aforementioned comparison provides some initial evidence for holding that there is a structural analogy between choosing a partner for procreative purposes and for choosing a fertilized egg cell after PGD. If this anaology is plausible, then it provides us with a reason for claiming that not all methods of genetic enhancement have to be morally problematic.

This argument assumes that the moral status of fertilized eggs is not yet a significant one. As it is unclear whether fertilized eggs metaphysically have a special moral status or not and there are significantly big interest groups in most Enlightened countries who affirm that both it has and it has not a significant moral status, I think that in this case the wonderful archievent of the norm of negative freedom ought to play the important part which implies that it ought to be up to the mother or parents in question to decide which moral status they attribute to their own fertilized eggs. In any case, I am aware that his question is a complex one, and I will deal with it in more detail in forthcoming publications.

Genetic Enhancement by means of Modification or Alteration and Classical Education: A Structural Analogy[7]

Habermas criticised the position that educational and genetic enhancements are parallel events (Habermas 2001, 91) whereby he referred to a position held by Robertson (1994, 167). I, on the other hand, wish to show that there is a structural analogy between educational and genetic enhancement such that the moral evaluation of these two procedures ought to be analogous, too (Habermas 2001, 87). I have already referred to the first parallel structure in the initial part of my article when I put forward reasons why it cannot be excluded that both procedures bring about changes which can be relevant for evolution.

Firstly, I wish to put forward some reasons why there could be parallels between these two procedures. Both procedures have in common that decisions are being made by parents concerning the development of their child, at a stage where the child cannot yet decide for himself what it should do. In the case of genetic enhancement we are faced with the choice between genetic roulette vs genetic enhancement. In the case of educational enhancement we face the options of a Kasper Hauser lifestyle vs parental guidance. Given these options, it seems most plausible to conclude that genetic enhancement and parental guidance usually bring about better results for the offspring than the other options mentioned, as the qualities brought about by means of enhancement are based upon parental choices which normally are made on the ground of experience. Parents usually love their children and want them to have the best starting point in life possible. Of course, it is not always the case that parental decisions bring about good results, but as a rule of thumb it should be possible to say that it is far more often the case that parental influence leads to better options than states which would have come about by chance or without any guidance. Parents uphold qualities on the basis of their experiences, and having experiences in the context of ethical decisions is necessary for making prudentially good decisions, as Aristotle has already remarked concerning the foundation of prudence. (NE 1142a)

Now, I will address two fundamental, but related claims which Habermas puts forward against the parallelisation of genetic and educative enhancement: Genetic enhancement is irreversible, and educative enhancement is revesible.

---

7   See Sorgner 2010, section 1.1.1.

Irreversibility of Genetic Enhancement

One claim against the parallelisation of genetic and educative enhancement is that genetic enhancement is irreversible, according to Habermas. However, this claim is implausible, if not false, as recent research has shown.

Let us consider the lesbian couple who are both deaf and who have chosen a deaf sperm donor to have a deaf child (see Agar 2004, 12-14). Actually, the child can hear a bit on one ear, but this is unimportant for my current purpose. According to the couple, deafness is not a defect, but it merely represents a being different. The couple was able to realise their wish and in this way managed to have a mostly deaf child. If germ line gene therapy worked, then they could have had a non deaf donor, changed the appropriate genes, and could have brought about a deaf child in this way. However, given the deafness in question is one of the inner ear. In that case, it would be possible for the person in question to later on go to the doctor and ask for a surgery in which he gets an implant, so that the person in question can hear. It is already possible to make such an operation and get such an implant.

Of course, it can get argued that in that case the genotype was not reversed but merely the phenotype. This is correct. However, the example also shows that qualities which came about due to a genetic setting are not irreversible. They can get changed by means of a surgery. Deaf people can undergo a surgery so that they can hear again, depending on the type of deafness they have or when the surgery takes place.

One could object that the consequences of educational enhancement can get reversed autonomously whereas in the case of genetic alterations one needs a surgeon or external help to bring about a reversed state. This is incorrect again, as I will show later. It is not the case that all consequences of educational enhancement can get reversed. In addition one can reply that by means of somatic gene therapy, it is even possible to change the genetic set up of a person. One of the most striking examples in this context is the siRNA therapy. By means of siRNA therapy, genes can get silenced. In the following paragraph, I state a brief summary of what siRNA therapy has achieved so far.

In 2002 the journal "Science" referred to RNAi as "Technology of the Year", and McCaffrey et al. published a paper in the journal "Nature" in which they specified that siRNA functions in mice and rats (2002, 38-9). That siRNA's can be used therapeutically in animals was published by Song et al. in 2003. By means of this type of therapy (RNA interference targeting Fas) mice can get protected from fulminant hepatitis (Song et al. 2003, 347-51). A year later it was shown that genes at transcriptional level can get silenced by means of siRNA (Morris 2004, 1289-1292). Due to the enormous potential of siRNA Andrew Fire and Craig Mello were awarded the noble prize in medicine for discovering RNAi mechanism in 2006.

Given the empirical data concerning siRNA, it is plausible to claim that theoretically the following process is possible and, hence, that genetic states do not have to be fixed: 1. An embryo with brown eyes can get selected by means of preimplantation genetic diagnosis – PGD; 2. The adult does not like his eye colour; 3. He asks medics to undergo siRNA therapy to change the gene related to his eye colour; 4. The altered genes bring it about that the eye colour changes.

Another option would be given, if germ line gene therapy worked which it does not so far. In that case, we could change a gene using germ line gene therapy to bring about a quality x. The quality x is disapproved of by the later adult. Hence, he decides to undergo siRNA therapy to silence the altered gene again. Such a procedure is theoretically possible. However, we do not have to use fictional examples to show that alterations brought about by genetic enhancement are reversible, but one must simply have a look at the latest developments in gene therapy.

The 23 year-old British male, Robert Johnson, suffered from Leber's congenital amaurosis which is an inherited blinding disease. Early in 2007, he had a surgery at Moorfields Eye Hospital and University College London's Institute of Ophthalmology which represented the world's first gene therapy trial for an inherited retinal disease. In April 2008 the "New England Journal of Medicine" published the results of this operation which revealed its success, as the patient has had a modest increase in vision afterwards, and no apparent side-effects (Maguire et al. 2008, 2240-2248).

In this case it was a therapeutic use of gene therapy. As genes can get altered for therapeutic purposes, it shows that they can get altered and that they could get altered for non therapeutic ends, too, if one wishes to uphold the problematic distinction between therapeutic and non therapeutic ends. The examples mentioned here clearly show that it is not the case that qualities brought about by means of genetic enhancement do not have to be irreversible. However, the parallels concerning genetic and educative enhancement go even further.

Reversibility of Educative Enhancement

According to Habermas, character traits brought about by educative means are reversible. Because of this assumption, which he must hold, he rejects that educative and genetic enhancement are parallel processes. Aristotle disagrees, and he is right in doing so. According to Aristotle, a *hexis*, a basic stable attitude gets established by means of repetition (EN 1103a). You become brave, if you continuously act in a brave manner. By playing a guitar, you turn into a guitar player. By acting with moderation, you become moderate. Aristotle makes clear that by means of repeating a certain type of action, you establish the type in your character, you form a basic stable attitude, a *hexis*. In the "Catagories", he makes clear that the *hexis* is extremely stable (Cat. 8, 8b27-35). In the "Nichomachean Ethics" he goes even further and claims that once one has established a basic stable attitude it is impossible to get rid of it again (EN III 7, 1114a19-21). Buddensiek has interpreted this passage correctly by pointing out that once a *hexis*, a basic stable attitude, was formed or established, it is an irreversible part of the character according to Aristotle (Buddensiek 2002, 190).

Aristotle's position gets support from Freud, who put forward the following claim: "It follows from what I have said that the neuroses can be completely prevented but are completely incurable" (Freud after Malcolm 1984, 24) whereby *Angstneurosen* were supposed to represent a particularly striking example (Rabelhofer 2006, 38).

A lot of time has passed since Freud and research has taken place. However, in recent publications concerning psychiatric and psychotherapeutic findings, it is still clear that psychological diseases can be incurable (Beese 2004, 20). Psychological disorders are not intentionally brought about by educative means. However, a lot of empirical research has been done in the field of illnesses, and their origin in early childhood. By showing that irreversible illnesses can come about due to events or actions which have taken place in childhood, it becomes clear that the same can happen by means of proper educative measures.

Medical research has shown and most physicians agree that PostTraumatic Stress Disorders can not only become chronic, but also lead to a permanent personality disturbance (Rentrop et al (eds.) 2009, 373). They come about due to exceptional events which represent an enormous burden and change within someone's life. Obsessional neuroses are another such case. According to the latest numbers, only 10 to 15 % of patients get cured, and in most cases it turns into a chronic disease (Rentrop et al (eds.) 2009, 368). Another disturbance which one could refer to is the borderline syndrome, which is a type of personality disorder. It can be related to events or actions which have taken place in early childhood, like violence or child abuse. In most cases this represents a chronic disease (Rentrop et al (eds.) 2009, 459).

Given the examples mentioned, it is clear that by means of actions and events which have taken place during ones lifetime, one can get into permanent and irreversible states. In the above cases, it is of disadvantage to the person in question. In the case of a Aristotelian *hexis*, it is an advantage for the person in question, given one establishes a virtue in this manner.

To provide further intuitive support for the position that qualities established by educational enhancement can be irreversible, one can simply think about ones learning to ride the bike, tie ones shoe laces, play the piano or speak ones mother tongue. Children get educated for years and years to undertake these tasks. Even when one moves into a different country or if one does not ride the bike for many years, it is difficult, if not impossible, to completely get rid of the capacity one has acquired before. Hence, it is very plausible that educative enhancement can have irreversible consequences, and Habermas is wrong twice. Genetic enhancement can have consequences which are reversible, and educative enhancement can have consequences which are irreversible. Given these insights, the parallelisation of genetic and educative enhancement gets additional support.

In a related unpublished paper, I deal with further reasons in favour of the parallel structure of genetic enhancement by selection and classical education, whereby I particularly focus on the questions of autonomy, instrumentalisation, equality, and the therapy/enhancement distinction.

Conclusion

Given the above analysis, I am already bound to conclude that Habermas is wrong concerning two fundamental issues when he denies that education and genetic enhancement by modification are parallel events. Given the plausibility of this structural analogy and the one between genetic enhancement by selection and choosing a partner for procreative purposes, it is plausible to hold that genetic enhancement does not have to be morally problematic.

In order to reach a clearer understanding of which types of genetic enhancement ought to be undertaken and which not, I would have to discuss the moral status of the embryo, and analyse which concept of the good ought to apply on a political level. I have not attempted to give a reply to both questions here.

What I wish to stress so far is that the norm of negative freedom is a precious achievement. During the Enlightenment, we have freed ourselves from the paternalistic oppression of religious and aristocratic leaders. Thereby, we have successfully established the right to live according to our own concept of the good, as long as it does not interfere with the rights of someone else, and I regard this archievement as a wonderful and extremely precious one which is worth fighting for. Consequently, I suggest that *In dubio pro libertate* as an adequate principle for a liberal-democracy. If there is a conflict between several groups of a certain, significant size, then I suggest that it usually ought to be the more liberal opinion which ought to be legalised. A state should also refrain from making normative demands based upon metaphysical and religious prejudices and rather consider both the latest scientific findings as well as the opinions which are widely being shared within ones own society. In the end a weighting of the following three

Evolution, Education, and Genetic Enhancement 97

pillars might be the most appropriate way of establishing a normative basis of a society: 1. Attitudes which are widely being shared ought to be considered; 2. Insights from the latest scientific research ought to be taken into account; 3. The wonderful norm of negative freedom always ought to have a central status and it ought to be taken into consideration that this norm gets undermined.

What are the most basic insights of this paper? As it is plausible to hold that structural analogies are given between various types of genetic enhancement and already known procedures which are mostly morally unproblematic, some reasons have been named which provide us with a reason for claiming that genetic enhancement does not have to be morally problematic. However, it is still an open question which procedures exactly can be described as morally adequate or appropriate.

Even if the parallelisation of educational and genetic enhancement is given, it does not solve the elemental challenges connected to it, like the question concerning the appropriate good as a basis for enhancement procedures. One important issue in this context was raised in a recent report. It summarises well the complexity and relevance of this issue which is the reason why I quote it in full length here. It provides a basis for further investigation and studies which have to follow the conclusion of this article:

> "Some have argued at least with regard to education that children possess a further right beyond health and safety. Article 26 of the United Nations Universal Declaration of Human Rights states that everyone has the right to education and that education shall be directed to the full development of the human personality and to the strengthening of respect for human rights and fundamental freedoms. These rights in turn suggest duties for parents and society. If education is a kind of social enhancement, this lays the groundwork for claiming that other kinds of enhancements might be the right of children and correlative duties of parents and children. Will, for example, children of the future be expected to receive enhancements of their bodies that lead to 'the full development of human personality'? Exactly what might be required will depend on the facts of the situation, of course. As we have said, context matters. But what this account shows is that there is at least a possible line of argument that supports not only the right of children to be enhanced, but also a duty of their parents or society to do this" (Allhoff et al. 2009, 32).

I already considered some social implications of the structural analogy between genetic enhancement by alteration and classical education within a talk and a public discussion which took place in October 2010 as part of the Bayreuther Dialoge. I have also mentioned some social implications concerning the structural analogy between genetic enhancement by selection and choosing a partner for procreative purposes in a talk I gave in October 2011 at Dublin City University. However, both reflections have not yet been made available in print. So far, I merely wish to have stated some reasons for not regarding genetic enhancement as necessarily morally problematic. I regard genetic enhancement as the most likely option for bringing about the posthuman. Consequently is is likely that genetic enhancement can also effect some processes of human evolution and natural selection receives further support from the process of human selection. Is this development necessarily a praiseworthy one? No, not necessarily, because there is always the risk that an invention has serious consequences. Only time will tell how to evaluate our further options concerning the enhancement of metahumans toward trans- and posthumans. However, I think that technologies in general have improved the quality of our lives significantly, I think that it is excellent that we have developed the various scientific methods for having all these choices, and I am very happy that I am living in a world with antiobiotics and where we have the option of being vaccinated which is clearly an enhancement technology. Hence, I can conclude that even though new dangers are connected to any innovation, so far our technological innovations have been extremely helpful for supporting the promoting the quality of human lives, and I expect that the same will be the case with respect to the various technologies conneted to the field of genetic enhancement.

Bibliography

Agar, N. (2004): Liberal Eugenics. In Defence of Human Enhancement. Blackwell Publishing, Oxford et al.
Allhoff, F. et al. (2009): Ethics of Human Enhancement. 25 Questions and Answers: http://www.humanenhance.com/NSF_report.pdf (20.03.2012; 4.30 pm).
Beese, F. (2004): Was ist Psychotherapie? Ein Leitfaden für Laien zur Information über ambulante und stationäre Psychotherapie. Vandenhoeck & Ruprecht, Goettingen.
Bostrom, N. (2005): Transhumanist Values. In: Review of Contemporary Philosophy 4: http://www.nickbostrom.com/ethics/values.pdf (20.03.2012, 4.40 pm).
Buddensiek, F. (2002): Hexis. In: Horn, C./Rapp, C. (2002): Wörterbuch der antiken Philosophie. Beck, Munich.
Habermas, J. (2001): Die Zukunft der menschlichen Natur. Auf dem Weg zu einer liberalen Eugenik? Suhrkamp Verlag, Frankfurt am Main.
Hayles, N.K. (1999): How we became Posthuman. Virtual bodies in Cybernetics, Literature, and Informatics. The University of Chicago Press, Chicago et al.
Haraway, D.J. (1991): Simians, Cyborgs, and women: The Reinvention of Nature. Free Association Books, London.
Japlonka, E./Lamb, M.J. (2005): Evolution in Four dimensions: Genetic, Epigenetic, Behavioral, and Symbolic Variation in the History of Life. The MIT Press, London et al.

Levenson, J.M./Sweatt J.D. (2005): Epigenetic mechanisms in memory formation. In: Nature Reviews Neuroscience 6, 108-118.
Malcolm, J. (1984): In the Freud Archives. Jonathan Cape Ltd, London.
Maguire, A.M. et al. (2008): Safety and efficacy of gene transfer for Leber's congenital amaurosis. In: New England Journal Medicine 358, 2240-2248.
McCaffrey, A.P. et al. (2002): RNA interference in adult mice. In: Nature 418, 38-39.
Morris, K. V. et al. (2004): Small Interfering RNA-Induced Transcriptional Gene Silencing in Human Cells. In: Science 305, 1289-1292.
Rabelhofer, B. (2006): Symptom, Sexualitaet, Trauma. Koenigshausen & Neumann, Wuerzburg.
Rentrop, M./Müller, R./Baeuml, J. (2009): Klinikleitfaden Psychiatrie und Psychotherapie. Elsevier, Urban & Fischer, Muenchen.
Robertson, J. A. (1994): Children of Choice. Freedom, and the New Reproductive Technologies. Princeton University Press, Princeton, NJ.
Sloterdijk, P. (2001): Nicht gerettet. Versuche nach Heidegger. Suhrkamp Verlag, Frankfurt am Main.
Song, E. et al. (2003): RNA interference targeting Fas protects mice from fulminant hepatitis. In: Nature Medicine 9, 347-351.
Sorgner, Stefan Lorenz (2009): Nietzsche, the Overhuman, and Transhumanism. In: Journal of Evolution and Technology. Vol. 20, Issue 1, March 2009, pgs 29-42.
Sorgner, Stefan Lorenz (2010): Beyond Humanism. Reflections on Trans- and Posthumanism. In: Journal of Evolution and Technology. Vol. 21, Issue 2, October 2010, 1-19.
Sorgner, Stefan Lorenz (2011): Zarathustra 2.0 and Beyond. Further Remarks on the Complex Relationship between Nietzsche and Transhumanism. In: The Agonist, Vol. IV, Iss. II, Fall 2011, 1-46.

On the Origins of Modern Science: Copernicus and Darwin

Francisco J. Ayala

Introduction

There is a version of the history of the ideas that sees a parallel between the Copernican and the Darwinian revolutions. In this view, the Copernican Revolution consisted in displacing the Earth from its previously accepted locus as the center of the universe, moving it to a subordinate place as just one more planet revolving around the sun. In congruous manner, the Darwinian Revolution is viewed as consisting of the displacement of humans from their exalted position as the center of life on earth, with all other species created for the service of humankind.

According to this version of intellectual history, Copernicus had accomplished his revolution with the heliocentric theory of the solar system. Darwin's achievement emerged from his theory of organic evolution.

Sigmund Freud refers to these two revolutions as "outrages" inflicted upon humankind's self-image and adds a third one, his own: "Humanity in the course of time had to endure from the hands of science two great outrages upon its naïve self-love. The first was when it realized that our earth was not the centre of the universe, but only a tiny speck in a world-system of a magnitude hardly conceivable; this is associated in our minds with the name of Copernicus, although Alexandrian doctrines taught something very similar. The second was when biological research robbed man of his peculiar privilege of having been specially created, and relegated him to a descent from the animal world, implying an ineradicable animal nature in him: this transvaluation has been accomplished in our own time upon the instigation of Charles Darwin, Wallace, and their predecessors, and not without the most violent opposition from their contemporaries. The third and most bitter blow upon man's craving for grandiosity" was meted out in the twentieth century by psychoanalysis, revealing that man's *ego* "is not even master in his own house."

What the standard version of the Copernican and Darwinian revolutions says is correct but inadequate, because it misses what is most important about these two intellectual revolutions, namely that they ushered in the beginning of science in the modern sense of the word. These two revolutions may jointly be seen as the one Scientific Revolution, with two stages, the Copernican and the Darwinian.

The Copernican Revolution

The Copernican Revolution was launched with the publication in 1543, the year of Nicolaus Copernicus' death, of his "De revolutionibus orbium celestium" (On the Revolutions of the Celestial Spheres), and bloomed with the publication in 1687 of Isaac Newton's "Philosophiae naturalis principia mathematica" (The Mathematical Principles of Natural Philosophy).

The discoveries by Copernicus, Kepler, Galileo, Newton, and others, in the sixteenth and seventeenth centuries, gradually ushered in a conception of the universe as matter in motion governed by natural laws. It was shown that Earth is not the center of the universe, but a small planet rotating around an average star; that the universe is immense in space and in time; and that the motions of the planets around the sun can be explained by the same simple laws that account for the motion of physical objects on our planet. Laws such as $f = m \times a$ (force = mass x acceleration); or the inverse-square law of attraction, $f = g(m_1 m_2)/r^2$ (the force of attraction between two bodies is directly proportional to their masses, but inversely related to the square of the distance between them).

These and other discoveries greatly expanded human knowledge. The conceptual revolution they brought about was more fundamental yet: a commitment to the postulate that the universe obeys immanent laws that account for natural phenomena. The workings of the universe were brought into the realm of science: explanation through natural laws. All physical phenomena could be accounted for as long as the causes were adequately known.

The Darwinian Revolution

The advances of physical science brought about by the Copernican Revolution had, however, driven mankind's conception of the universe to a split-personality state of affairs, which persisted well into the mid-nineteenth century. Scientific explanations, derived from natural laws, dominated the world of nonliving matter, on the Earth as well as in the heavens. However, supernatural explanations, which depended on the unfathomable deeds of the Creator, were accepted as explanations of the origin and configuration of living creatures.

Authors, such as William Paley (1743-1805), argued that the complex design of organisms could not have come about by chance, or by the mechanical laws of physics, chemistry, and astronomy, but was rather accomplished by an Omniscient and Omnipotent Deity, just as the complexity of a watch, designed to tell time, was accomplished by an intelligent watchmaker.

It was Darwin's genius to resolve this conceptual quagmire. Darwin completed the Copernican Revolution by drawing out for biology the notion of nature as a lawful system of matter in motion that human reason can explain without recourse to supernatural agencies.

The conundrum faced by Darwin can hardly be overestimated. The strength of the argument from design to demonstrate the role of the Creator had been forcefully set forth by Paley, as well as by other philosophers and theologians. Wherever there is function or design, we look for its author. It was Darwin's greatest accomplishment to show that the complex organization and functionality of living beings can be explained as the result of a natural process – natural selection – without any need to resort to a Creator or other external agent. The origin and adaptations of organisms in their profusion and wondrous variations were thus brought into the realm of science.

Darwin's Theory

Darwin considered natural selection, rather than his demonstration of evolution, his most important discovery, and designated it as "my theory", a designation he never used when referring to the evolution of organisms. The discovery of natural selection, Darwin's awareness that it was a greatly significant discovery because it was science's answer to Paley's argument from design, and Darwin's designation of natural selection as "my theory" can be traced in Darwin's "Red Notebook" and "Transmutation Notebooks B to E", which he started in March 1837, not long after returning (on October 2, 1836) from his five-year voyage on the *Beagle*, and completed in late 1839.

Early in the Notebooks, Darwin registers his discovery of natural selection and repeatedly refers to it as "my theory." From then until his death in 1882, Darwin's life would be dedicated to substantiating natural selection and its companion postulates, mainly the pervasiveness of hereditary variation and the enormous fertility of organisms, which much surpassed the capacity of available resources. He relentlessly pursued observations and performed experiments in order to test the theory and resolve presumptive objections.

The evolution of organisms was commonly accepted by naturalists in the middle decades of the nineteenth century. The distribution of exotic species in South America, in the Galápagos Islands, and elsewhere, and the discovery of fossil remains of long-extinguished animals, confirmed the reality of evolution in Darwin's mind. The intellectual challenge was to explain the origin of distinct species of organisms, how new ones adapted to their environments, that "mystery of mysteries", as it had been labeled by Darwin's older contemporary, the prominent scientist and philosopher Sir John Herschel (1792-1871).

In his "Autobiography", Darwin wrote, "The old argument of design in nature, as given by Paley, which formerly seemed to me so conclusive, falls, now that the law of natural selection has been discovered. We can no longer argue that, for instance, the beautiful hinge of a bivalve shell must have been made by an intelligent being, like the hinge of a door by a man."

Wherever there is function or design, we look for its author. Whenever we see a watch, we know that there is a watchmaker. Similarly, the structures, organs and behaviors of living beings are directly organized to serve certain purposes or functions. The functional design of organisms and their features would therefore seem to argue for the existence of a Designer. Darwin's great accomplishment was to show that the diversity and adaptations of living beings can be explained as the result of natural selection.

Darwin accepted that organisms are "designed" for certain purposes, that is, they are functionally organized. Organisms are adapted to certain ways of life and their parts are adapted to perform certain functions. Fish are adapted to live in water, kidneys are designed to regulate the composition of blood, the human hand is made for grasping. But Darwin went on to provide a natural explanation of the design. The seemingly purposeful aspects of living beings could now be explained, like the phenomena of the inanimate world, by the methods of science, as the result of natural laws manifested in natural processes.

Darwin's Explanation of Design

Darwin is deservedly given credit for the theory of evolution. The evolution of organisms was commonly accepted by naturalists in the middle decades of the 19th century. In "The Origin of Species", Darwin accumulated overwhelming evidence demonstrating the evolution of organisms. However, Darwin accomplished something much more important for intellectual history than demonstrating evolution. Indeed, accumulating evidence for common descent with diversification may very well have been a subsidiary objective of Darwin's masterpiece. Darwin's "The Origin of Species" is, first and foremost, a sustained argument to solve the problem of how to account scientifically for the design of organisms. Darwin brings about the evidence for evolution because evolution is a necessary consequence of his theory of design. Darwin's most revolutionary achievement is that he extended the Copernican revolution to the world of living things.

Alfred Russel Wallace

Alfred Russel Wallace (1823-1913) is famously given credit for discovering, independently of Darwin, natural selection as the process accounting for the evolution of species. On June 18, 1858, Darwin wrote to Charles Lyell that he had received by mail a short essay from Wallace such that "if Wallace had my [manuscript] sketch written in [1844] he could not have made a better abstract." Darwin was thunderstruck.

Wallace's independent discovery of natural selection is remarkable. Wallace, however, was not interested in explaining design, but rather in accounting for the evolution of species, which he saw as a sustained and progressive process, as indicated in his paper's title: "On the Tendency of Varieties to Depart Indefinitely from the Original Type." Wallace thought that evolution proceeds indefinitely and is progressive. Darwin, on the contrary, did not accept that evolution would necessarily represent progress or advancement, nor did he believe that evolution would always result in morphological change over time; rather, he knew of the existence of "living fossils", organisms that had remained unchanged for millions of years. For example, "some of the most ancient Silurian animals, as the Nautilus, Lingula, etc., do not differ much from living species." (We now know that the Silurian geological period lasted from 444 to 416 million years ago.)

In 1858, Darwin was at work on a multivolume treatise, intended to be titled "On Natural Selection". Wallace's paper stimulated Darwin to write "The Origin of Species", which would be published the following year. Darwin intended this as an abbreviated version of the much longer book he had intended to write. As I have noted earlier, Darwin's focus, in "The Origin of Species" as elsewhere, was the explanation of design, with evolution playing the subsidiary role of supporting evidence.

Darwin's "Origin"

"The Origin of Species" is usually characterized as the most important book ever published about the theory of evolution. This is correct, but not so much because of the numerous observations and facts magisterially gathered by Darwin demonstrating the evolution of organisms, but rather because with his theory of natural selection he advanced an explanation that accounted not only for the evolution of organisms, but also for their adaptations. The Introduction and Chapters I through VIII of "The Origin of Species" explain how natural selection accounts for the adaptations and behaviors of organisms, their "design." The extended argument starts in Chapter I, where Darwin describes the successful selection of domestic plants and animals and, with considerable detail, the success of pigeon fanciers seeking exotic "sports". The success of plant and animal breeders manifests how much selection can accomplish by taking advantage of spontaneous hereditary variations that occur in organisms but happen to fit the breeders' objectives. A sport (mutation) that first appears in an individual can be multiplied by selective breeding, so that after a few generations that sport becomes fixed in a breed, or "race". The familiar breeds of dogs, cattle, chickens, and food plants have been obtained by this process of selection practiced by people with particular objectives.

The ensuing chapters (II-VIII) of "The Origin of Species" extend the argument to variations propagated by natural selection for the benefit of the organisms themselves, rather than by artificial selection of traits desired by humans. As a consequence of natural selection, organisms exhibit design, that is, exhibit adaptive organs and functions. The design of organisms as they exist in nature, however, is not "intelligent design", imposed by God as a Supreme Engineer or by humans; rather, it is the result of a natural process of selection, promoting the adaptation of organisms to their environments. This is how natural selection works: Individuals that have beneficial variations, that is, variations that improve their probability of survival and reproduction, leave more descendants than individuals of the same species that have less beneficial variations. The beneficial variations will consequently increase in frequency over the generations; less beneficial or harmful variations will be eliminated from the species. Eventually, all individuals of the species will have the beneficial features; new features will arise over eons of time.

Organisms exhibit complex design, but this design does not exhibit, in current language, "irreducible complexity", emerging all of a sudden in full bloom. Rather, according to Darwin's theory of natural selection, the design has arisen gradually and cumulatively, step by step, promoted by the reproductive success of individuals with incrementally more adaptive elaborations.

It follows from Darwin's explanation of adaptation that evolution must necessarily occur as a consequence of organisms becoming adapted to different environments in different localities, and to the ever-changing conditions of the environment over time; and as hereditary variations become available at a particular time that improve, in that place and at that time, the organisms' chances of survival and reproduction. "The Origin of Species"' evidence for biological evolution is central to Darwin's explanation of design, because this explanation implies that biological evolution occurs, which Darwin therefore seeks to demonstrate in most of the remainder of the book (chapters IX–XIII. In the sixth edition of "The Origin of Species", these are chapters X-XIV, because Darwin had added a new chapter VII: "Miscellaneous objections to the theory of natural selection").

In the concluding Chapter XIV of "Origin", Darwin returns to the dominant theme of adaptation and design. In an eloquent final paragraph, Darwin asserts the "grandeur" of his vision:

> "It is interesting to contemplate an entangled bank, clothed with many plants of many kinds, with birds singing on the bushes, with various insects flitting about, and with worms crawling through the damp earth, and to reflect that these *elaborately constructed* forms, *so different* from each other, and dependent on each other *in so complex a manner*, have all been produced by laws acting around us. ... Thus, from the war of nature, from famine and death, the most exalted object which we are capable of conceiving, namely, the production of the higher animals, directly follows. There is grandeur in this view of life, with its several powers, having been originally breathed into a few forms or into one; and that, whilst this planet has gone cycling on according to the fixed law of gravity, from so simple a beginning endless forms most beautiful and most wonderful have been, and are being, evolved" (emphasis added).

Adaptation and Evolution

Darwin's "The Origin of Species" addresses the issue of how to account for the adaptive configuration of organisms and their parts, which are so obviously designed to fulfill certain functions. Darwin argues that hereditary adaptive variations ("variations useful in some way to each being") occasionally appear, and that these are likely to increase the reproductive chances of their carriers. The success of pigeon fanciers and animal breeders clearly shows the occasional occurrence of useful hereditary variations. In nature, over the generations, Darwin's argument continues, favorable variations will be preserved, multiplied, and conjoined; injurious ones will be eliminated. Evolution affects all aspects of an organism's life – morphology (form and structure), physiology (function), behavior, and ecology (interaction with the environment). Underlying these changes are changes in the hereditary materials. Hence, in genetic terms, evolution consists of changes in the organisms' hereditary makeup. Darwin formulated natural selection primarily as differential survival. The modern understanding of the principle of natural selection is formulated in genetic and statistical terms as differential reproduction. Natural selection simply implies that some genes and genetic combinations are transmitted to the following generations more frequently than their alternates. Favored genes will become more common in every subsequent generation, and their alternates less common. Natural selection is a statistical bias in the relative rate of reproduction of alternative genes.

Evolution can be seen as a two-step process. First, hereditary variation arises by mutation; second, selection occurs by which useful variations increase in frequency and those that are less useful or injurious are eliminated over the generations. "Useful" and "injurious" are terms used by Darwin in his definition of natural selection. The significant point is that individuals having useful variations "would have the best chance of surviving and procreating their kind". As a consequence, useful variations increase in frequency over the generations, at the expense of those that are less useful or injurious.

Natural selection is much more than a "purifying" process, for it is able to generate novelty by increasing the probability of otherwise extremely improbable genetic combinations. Natural selection in combination with mutation becomes, in this respect, a creative process. Moreover, it is a process that has been occurring for many millions of years, in many different evolutionary lineages and a multitude of species, each consisting of a large number of individuals. Evolution by mutation and natural selection has produced the enormous diversity of the living world with its wondrous adaptations. Several hundred million generations separate modern animals from the early animals of the Cambrian geological period (542 million years ago). The number of mutations that can be tested, and those eventually selected, in millions of individual animals over millions of generations is difficult for a human mind to fathom, but we can readily understand that the accumulation of millions of small, functionally advantageous changes could yield remarkably complex and adaptive organs, such as the eye.

Natural selection does not operate as a sieve that retains the rarely arising useful genes and lets go the more frequently arising harmful mutants; at least, not only. Natural selection acts in the filtering way of a sieve, but it is much more than a purely negative process, for it is able to generate novelty by increasing the probability of otherwise extremely improbable genetic combinations. Natural selection is thus a creative process. It does not "create" the entities (mutations) upon which it operates, but it produces adaptive (functional) genetic combinations that could not have existed otherwise.

Darwin's Critics

Critics have sometimes alleged as evidence against Darwin's theory of evolution examples showing that random processes cannot yield meaningful, organized outcomes. It is thus pointed out that a series of monkeys randomly striking letters on a typewriter would never write "The Origin of Species", even if we allow for millions of years and many generations of monkeys pounding at typewriters. This criticism would be valid if evolution would depend only on random processes. But natural selection is a non-random process that promotes adaptation by selecting combinations that "make sense", i.e., that are useful to the organisms. The analogy of the monkeys would be more appropriate if a process existed by which, first, meaningful words would be chosen every time they appeared on the typewriter; and then there would also be typewriters with previously selected words rather than just letters in the keys, and again there would be a process to select meaningful sentences every time they appeared in this second typewriter. If every time words such as "the", "origin", "species", and so on, appeared in the first kind of typewriter, they each became a key in the second kind of typewriter, meaningful sentences would occasionally be produced in this second typewriter. If such sentences became incorporated into keys of a third type of typewriter, in which meaningful paragraphs were selected whenever they appeared, it is clear that pages and even chapters "making sense" would eventually be produced. The end product would be an "irreducibly complex" text.

We need not carry the analogy too far, since the analogy is not fully satisfactory, but the point is clear. Evolution is not the outcome of purely random processes, but rather there is a "selecting" process, which picks up adaptive combinations because these reproduce more effectively and thus become established in populations. These adaptive combinations constitute, in turn, new levels of organization upon which the mutation (random) plus selection (non-random or directional) process again operates. The complexity of organization of animals and plants is "irreducible" to simpler components in one or very few steps, but not through the millions and millions of generations and the multiplicity of steps and levels made possible by eons of time.

A critical point is that evolution by natural selection is an incremental process, operating over eons of time and yielding organisms better able to survive and reproduce than others, which typically differ from one another at any one time only in small ways; for example, the difference between producing more or fewer progeny or between having or lacking an enzyme able to catalyze the synthesis of one particular amino acid. Notice also that increased complexity is not a necessary outcome of natural selection, although complexity increases from time to time in some lineages of descent, so that, although rare, these lineages are very conspicuous over time's eons. That is, increased complexity is not a necessary consequence of evolution by natural selection, but rather emerges occasionally. The longest living organisms on Earth are the microscopic bacteria, which have continuously existed on our planet for three and a half billion years and yet exhibit no greater complexity than their old time ancestors. More complex organisms came about much later, without the elimination of their simpler relatives. For example, the primates appeared on earth some fifty million years ago and our species, *Homo sapiens*, came about two hundred thousand years ago.

Step-by-Step

Individuals of a given species differ from one another at any one time only in small ways; for example, the difference between bacteria that have or lack an enzyme able to synthesize the sugar lactose or between moths that have light or dark wings. These differences typically involve one or only a few genes, but they can make the difference between survival or death, as in the evolution of resistance to DDT in disease-transmitting mosquitoes or to antibiotics in people. Consider a different sort of example. Some pocket mice (*Chaetodipus intermedius*) live in rocky outcrops in Arizona. Light, sandy-colored mice are found in light-colored habitats, whereas dark (melanic) mice prevail in dark rocks formed from ancient flows of basaltic lava. The match between background and fur color protects the mice from avian and mammal predators that hunt guided largely by vision. Mutations in one single gene (coding for the melanocortin-1-receptor, represented as *MC1R*) account for the difference between light and dark pelage (Nachman et al. 2003).

Adaptations that involve complex structures, functions, or behaviors involve numerous genes. Many familiar mammals, but not marsupials, have a placenta. Marsupials include the familiar kangaroo and other mammals native primarily to Australia and South America. Dogs, cats, mice, donkeys, and primates are placental. The placenta makes it possible to extend the time the developing embryo is kept inside the mother and thus make the newborn better prepared for independent survival. However, the placenta requires complex adaptations, such as the suppression of harmful immune interactions between mother and embryo, delivery of suitable nutrients and oxygen to the embryo, and the disposal of embryonic wastes. The mammalian placenta evolved more than 100 million years ago and proved a successful adaptation, contributing to the explosive diversification of placental mammals in the Old World and North America.

The placenta also has evolved in some fish groups, such as *Poeciliopsis*. Some *Poeciliopsis* species hatch eggs. The females supply the yolk in the egg, which furnishes nutrients to the developing embryo (as in chicken). Other *Poeciliopsis* species, however, have evolved a placenta through which the mother provides nutrients to the developing embryo. Molecular biology has made possible the reconstruction of the evolutionary history of *Poeciliopsis* species. A surprising result is that the placenta evolved independently three times in this fish group. The required complex adaptations accumulated in each case in less than 750,000 years (Reznick et al. 2002; Avise 2006).

Natural selection produces combinations of genes that would seem highly improbable because natural selection proceeds stepwise over long periods of time. Consider the evolution of the eye in humans and other vertebrates. Perception of light, and later vision, were important for the survival and reproductive success of their ancestors, because sunlight is a predominant feature of the environment. Accordingly, natural selection favored genes and gene combinations that increased the functional efficiency of the eye. Such mutations gradually accumulated, eventually leading to the highly complex and efficient vertebrate eye (Ayala 2007).

Chance and Necessity

Engineers have a preconception of what the design of a contrivance of structure is supposed to achieve, and will select suitable materials and arrange them in a preconceived manner so that it fulfills the intended function. On the contrary, natural selection has no foresight, nor does it operate according to some preconceived plan. Rather it is a purely natural process resulting from the interacting properties of physicochemical and biological entities. Natural selection is simply a consequence of the differential multiplication of living beings, as pointed out. It has some appearance of purposefulness because it is conditioned by the environment: which organisms reproduce more effectively depends on what variations they possess that are useful in the place and at the time where the organisms live. But natural selection does not anticipate the environments of the future; drastic environmental changes may be insuperable to organisms that were previously thriving. Species extinction is the common outcome of the evolutionary process. The species existing today represent the balance between the origin of new species and their eventual extinction. More than 99 percent of all species that ever lived on Earth have become extinct without issue. These may have been more than one billion species; the available inventory of living species has identified and described less than two million out of some ten million estimated to be now in existence.

The team of typing monkeys is a bad analogy of evolution by natural selection, because it assumes that there is "somebody" who selects letter combinations and word combinations that make sense. In evolution there is no one selecting adaptive combinations. These select themselves because they multiply more effectively than less adaptive ones.

The process of natural selection can explain the adaptive organization of organisms, as well as their diversity and evolution, as a consequence of their adaptation to the multifarious and ever changing conditions of life. The fossil record shows that life has evolved in a haphazard fashion. The radiations, expansions, relays of one form by another, occasional but irregular trends, and the ever-present extinctions, are best explained by natural selection of organisms subject to the vagaries of genetic mutation and environmental challenge. The scientific account of these events does not necessitate recourse to a preordained plan, whether imprinted from without by an omniscient and all-powerful Designer, or resulting from some immanent force driving the process towards definite outcomes. Biological evolution differs from a painting or an artifact in that it is not the outcome of preconceived design.

Natural selection accounts for the "design" of organisms, because adaptive variations tend to increase the probability of survival and reproduction of their carriers at the expense of maladaptive, or less adaptive, variations. The arguments of Paley against the incredible improbability of chance accounts of the adaptations of organisms are well taken as far as they go. But neither Paley nor any other author before Darwin was able to discern that there is a natural process (namely, natural selection) that is not random, but rather is oriented and able to generate order or "create." The traits that organisms acquire in their evolutionary histories are not fortuitous but determined by their functional utility to the organisms, "designed" as it were to serve their life needs.

Chance is, nevertheless, an integral part of the evolutionary process. The mutations that yield the hereditary variations available to natural selection arise at random, independently of whether they are beneficial or harmful to their carriers. But this random process (as well as others that come to play in the great theatre of life) is counteracted by natural selection, which preserves and multiplies what is useful and eliminates the harmful. Without hereditary mutation, evolution could not happen because there would be no variations that could be differentially conveyed from one to another generation. But without natural selection, the mutation process would yield disorganization and extinction because most mutations are disadvantageous. Mutation and selection have jointly driven the marvelous process that starting from microscopic organisms has yielded orchids, birds, and humans.

The theory of evolution conveys chance and necessity jointly intricated in the stuff of life; randomness and determinism interlocked in a natural process that has spurted the most complex, diverse, and beautiful entities in the universe: the organisms that populate the earth, including humans who think and love, endowed with free will and creative powers, and able to analyze the process of evolution itself that brought them into existence. This is Darwin's fundamental discovery, that there is a process that is creative though not conscious, a process that creates design without necessitating a designer.

Coda

Darwin completed the Copernican Revolution. The most important significance in the history of ideas of the Copernican and the Darwinian revolutions is that they jointly brought in the emergence of science, as a way of explaining all phenomena of the natural world. Science had, thus, come of age and, with science, modern technology, which overwhelmingly is the dominant feature of the world in which we live.

Bibliography

Avise, J.C. (2006): Evolutionary Pathways in Nature. A Phylogenetic Approach. Cambridge University Press, Cambridge et al., UK.

Ayala, F.J. (2007): Darwin's greatest discovery: Design without designer. In: Proceedings of the National Academy of Sciences, USA 104, 8567-8573.

Nachman, M.W./Hoekstra, H.E./D'Agostino, S.L. (2003): The genetic basis of adaptive melanism in pocket mice. In: Proceedings of the National Academy of Sciences, USA 100, 5268-5273.

Reznick, D.N./Mateos, M./Springer, M.S. (2002): Independent Origins and Rapid Evolution of the Placenta in the Fish Genus Poeciliopsis. In: Science 298, 1018-1020.

Technology as a New Theology From "New Atheism" to Technotheism

Mikhail Epstein

By the late 20th century, Western science and religion had established a relationship of mutual respect and tolerance, which was partially conditioned by the evidently catastrophic consequences of militant atheism in Communist countries. However in the post-Communist world, the historical lesson of atheism, destructive towards spirituality and culture, began to be obliterated.This made easier for some representatives of contemporary science to mount an offensive against religion once again, not against any specific religions but against the essence of religion as such.

My essay begins with the critique of Richard Dawkins' views as exposed in his book "The God Delusion" (2006) and further proceeds toward arguments for the existence of the Supreme Mind based on the advancements of contemporary science and technology: the position that I define as "technotheism."

Atheism and Genocentrism: The Morality of Genes

Dawkins is the famous evolutionary biologist and popular science writer, and his book is considered to be the most consistent manifesto of "the New Atheism", allegedly representing the scientific world-view.[1] This brand of atheism is "new" mainly concerning the time of its origin, whereas when it comes to the essence of its arguments, it is reminiscent of the 18th and 19th centuries rationalism and positivism.

The New Atheists are critical of the social-historical aspects of religion, referring to the latest manifestations of religious fundamentalism and fanaticism. And not only of the Islamic variety. The growing disappointment with Bush's Protestant-political brand of Messianism, which has resulted in heavy losses

---

1 Dawkins, Richard. The God Delusion. Boston, New York: A Mariner Book, 2006. Dennett, Daniel. Breaking the Spell: Religion As a Natural Phenomenon. Penguin Books, 2007. Harris, Sam. Letter to a Christian Nation.Vintage, 2008. Hitchens, Christopher. God Is Not Great: The Case Against Religion.Atlantic Books, 2007. These four authors are referred to as "the four riders of New Atheism"; naturally, this is an allusion to the four riders of the Apocalypse.

sustained during the Iraq war and the loss of USA's prestige throughout the world, has also contributed to a demand for fresh approaches to religion. According to the ICM survey of 2004, 91% of Americans believe in the supernatural, 74% believe in the afterlife, while 71% of them are prepared to die for their faith. Taking this into account, one can understand Richard Dawkins's complaint that, at the beginning of the 21st century, America still lives in an epoch of "theocratic Middle Ages". "The genie of religious fanaticism is rampant in present-day America, and the Founding Fathers would have been horrified."[2] (Dawkins 2008, 63)

According to Dawkins, religion is incompatible not only with science but also with morality. It is not only a great lie but also a great evil, and every individual who allows the possibility of religious beliefs being true only contributes to the increase of social evil, inequality, cruelty, militarism, etc. Generally, God as a person, the main character and the inspiration behind the Bible, does not arouse benevolent feelings in Dawkins. He refers to God as "arguably the most unpleasant character in all fiction: jealous and proud of it; a petty, unjust, unforgiving control-frak; a vindictive, bloodthirsty ethnic cleanser, a misogynistic, homophobic, racist, infanticidal, genocidal, filicidal, pestilential, megalomaniacal, sadomasochistic, capriciously malevolent bully" (Dawkins 2008, 49), "psychotic delinquent" (Dawkins 2008, 59). But the most important fact here is not that God is bad but that He, thank God, does not exist. That is why Dawkins magnanimously rejects the issue of God's personal lack of morality. "I am not attacking the particular qualities of Yahweh, or Jesus, or Allah ..." (Dawkins 2008, 52). Such attacks are not worth the trouble, for "God, in the sense defined, is a delusion." (Dawkins 2008, 52).

The theoretical basis of Dawkins's atheism is absolutely canonical, one could even say – of the "Old Ritual" type, bearing in mind that atheism has its own dogmatic tradition. It is a denial of any kind of reality beyond the scope of matter, given to us through our senses. I shall quote Dawkins's main thesis (which, incidentally, quite astonishingly corresponds to V. I. Lenin's book "Materialism and Empiriocriticism", 1909):

"Human thoughts and emotions *emerge* from exceedingly complex interconnections of physical entities within the brain. An atheist in this sense of philosophical naturalist is somebody who believes there is nothing beyond the natural, physical world, no *super*natural creative intelligence lurking behind the observable universe, no soul that outlasts the body and no miracles – except in the sense of natural phenomena that we don't yet understand. If there is something that appears to lie beyond the natural world as it is now imperfectly understood, we hope eventually to understand it and embrace it within the natural. As ever when we unweave a rainbow, it will not become less wonderful." (Dawkins 2008, 34-35)

---

2   Dawkins, R. (2006): The God Delusion. A Mariner Book. Boston, New York.

What is the atheistic morality that Dawkins opposes to the religious morality, basing it on a supposedly scientific-evolutionary world-view? On the biological level, the basis of the entire material universe in Dawkins is the selfish gene, which passionately strives to multiply. Starting from his book "The Selfish Gene" (1976), Dawkins has steadfastly preached the morality of genes as masters of the living universe; within the framework of this world-view, organisms and populations are mere slaves to genes, or more precisely, their copying machines. To the question of what came first – the hen or the egg, Dawkins gives a precise answer: the hen is merely a means of multiplying eggs. Genes create the mechanism of natural selection in order to multiply more effectively. Whatever lives in this world, all its creatures, organisms, peoples, history, culture, persons, passions, geniuses, masterpieces, religion – all of the above is merely a curtain behind which there unfolds the struggle of genes to master the world. In Dawkins, *genocentrism* essentially turns out to be much more reductive than mediaeval theocentrism, whose sway over America he is condemning. And yet, God does love His creations, and has made one of them, man, the crown of the universe; the world is not entirely reduced to its First Cause, it has its own value: to God, a human being is a goal, not a means. Genes are indifferent to the persons they inhabit and use them as a mere weapon in the course of their struggle, heartlessly trashing them the moment they lose their reproductive value.

According to Dawkins, the laws of man's morality are also determined by the gene whose egoism utilizes even the altruism of certain organisms. They sacrifice themselves for the sake of their closest relatives – so that the genes would have better conditions for multiplying through these relatives. "A gene that programs individual organisms to favour their genetic kin is statistically likely to benefit copies of itself. Such a gene's frequency can increase in the gene pool to the point where kin altruism becomes the norm." (Dawkins 2008, 247). Well, Dawkins manages to explain why by helping our relatives, by sacrificing for them, we still bear in mind the interests of our common genes (even though it is not quite clear why a developed individual should place these above his/her own interests). But why should one behave altruistically with non-relatives, with aliens? What interest do my genes have in my organism, their carrier, doing other people good – and not pillage, rape, kill with a view to multiplying its genes?

"In ancestral times, we had the opportunity to be altruistic only towards close kin and potential reciprocators. Nowadays that restriction is no longer there, but the rule of thumb persists. Why would it not? It is just like sexual desire. We can no more help ourselves feeling pity when we see a weeping unfortunate (who is unrelated and unable to reciprocate) than we can help ourselves feeling lust for a member of the opposite sex (who may be infertile or otherwise unable to reproduce). Both are misfirings, Darwinian mistakes: blessed, precious mistakes." (Dawkins 2008, 253).

Even Lenin's materialism did not sink to such a level of reductionism. Dawkins believes that the only purpose of evolution is reproduction. That is why sexual passion as such, without being linked to fertilisation, is an evolutionary "misfiring". What is also erroneous is that billions of people couple with those they love and desire not for the purpose of begetting offspring, but for the sake of emotional and sexual satisfaction (if we are still to adhere to the language of materialism). What comes to the fore in Dawkins here is the most ascetic and conservative Catholic doctrine, preaching that spousal intimacy is justified only by the objective of conception, whereas in all other cases it is merely sinful. All major Christian confessions, including Catholicism and Orthodoxy, have already renounced such a narrowly utilitarian approach to the mystery of marriage, pointing out that spousal intimacy in itself, apart from the purpose of child bearing, has a spiritual and moral value: "that two should be one flesh". But in Dawkins's view, Darwin's evolution is incomparably more cruel, ascetic and hypocritical than God in the views of Jews, Christians, and Muslims. The objectives of this evolution match only when it comes to sexual intimacy for the purpose of conception. Everything else is an error, a deviation from the right evolutionary path. If you feel passion towards a beautiful woman, even your own wife, and do not set yourself the task of childbearing, you should be aware of the fact that you are erring against the all-powerful and demanding Evolution.

No God, not even the "vindictive despot" from the Old Testament, so odious to Dawkins, can compare to the Selfish Gene in terms of self-interest and self-love, or to Evolution when it comes to ruthlessness and pragmatism. God loved His people and released them from slavery, taught people to love not only Himself but their neighbours; he also taught them not to kill, steal, swear falsely, abduct the women of others or take the property of others. The Selfish Gene, however, would allow all this, and would even demand of people that they kill, steal and rape if should it be necessary for its self-propagation. Evolution, on the other hand, would reject all feelings, aspirations, motives of love and good unless they directly contribute to the goal of evolution. Evolutionism may turn into dangerous fanaticism. We know how the dogmatism of all ideas based on unmasking "religious delusions" gradually intensifies. First, there come enlighteners, then Marxists, followed by Leninists, then Stalinists – and look, thousands and millions of heads infected with harmful "delusions" are rolling already.

And why, in fact, does Dawkins call the errors like useless sexual passion and the altruistic desire to help others "precious" and "wonderful"? How can a genuine scientist, an evolutionist to boot, value false behaviour, at odds with the goals of evolution, so highly? Is this not a poor justification of the above: "Do not, for one moment, think of such Darwinizing as demeaning or reductive of the noble emotions of compassion and generosity. Nor of sexual desire. Sexual desire, when channelled through the conduits of linguistic culture, emerges as great poetry and drama: John Donn's love poems, say, or *Romeo and Julie.* " (Dawkins 2008, 253) Ah, how touching! But what do we need Shakespeare and Donne for if they are of no use to Genes? Throw them to the garbage dump of Evolution! Dawkins's apologetic gesture, his attempt to explain the usefulness of sexual passion and altruistic tendencies by claiming that they provide a basis for creating great poetry, cannot elicit anything but cheery ridicule from both "idealists" and "materialists" alike. From the point of view of idealists, love and good do not require justification in verse, for they are precious in themselves, even more precious than verses. As for materialists, no verses can serve to justify errors against nature, for verses themselves are even more erroneous and useless. Therefore, the only thing Dawkins's compromise can elicit is contempt from both sides.

This apologetic gesture is merely indicative of a lack of consistency on the part of a hesitant atheist, a sentimental introduction to a truly evolutionary morality, which will not allow such deviations from the general line of Evolution. "Sexual lust is the driving force behind a large proportion of human ambition and struggle, and much of it constitutes a misfiring. There is no reason why the same should not be true of the lust to be generous and compassionate, if this is the misfired consequence of ancestral village life... Those rules still influence us today, even where circumstances make them inappropriate to their original functions." (Dawkins 2008, 254) Why, if they are inappropriate, to hell with those patterns! We can do without them. Onward, along the bright path of Evolution, under the management of the all-powerful Genes!

Tell the entire society that "the lust to be generous and compassionate" is the erroneous consequence of primitive customs, no longer suitable for contemporary civilization. And let society be well aware of, believe and live according to Dawkins's maxim. In twenty years, society will explode as a result of internal arguments and "non-compassion", and there will be no one left to multiply genes.

Where Dawkins's genetic-evolutionary atheism is consistent, it is immoral; and where it is moral, it is inconsistent, stealthily introducing morality as an error, however wonderful it may be. Whatever Dawkins's atheism uses to explain human, creative, moral, spiritual motives and aims of human behaviour, he takes it from traditional ethics, stemming from religious world-views, and what he adds himself is a combination of reductionism and anthropomorphism. Dawkins reduces man's personality to the elements making up the human organism, such as genes and chromosomes, which are reputedly "human" in themselves, irrespective of man himself. Genes are "selfish", they set their own goals – this is a kind of pseudo-scientific mythology, much more superstitious than the Old or New Testament that Dawkins critically "demythologizes". In order to derive the major from the minor, such atheism provides the minor with the characteristics of the major. Aspirations towards a totally scientific outlook turn into grotesque tomfoolery when a particle of matter is endowed with will, reason and moral motivation. Atheism complains that religion alienates man's qualities (creation, reason, love, morality), turning them over to a Higher Being. But atheism itself also alienates them from man and turns them over to genes or elementary particles of physical matter.

Information Technologies and an Argument for the Existence of the Supreme Mind

I shall now switch from a critique of the latest form of atheism to expounding on a position that appears constructive to me. The logic of scientific-technical progress suggests to us, on the evidence of our own increasing mastery of the world, that the Universe has a Maker. It is much easier for us today, on the basis of contemporary scientific data, to believe in the Supreme Mind than it was for our less knowledgeable ancestors. I use the word argument, not proof, because the existence of God, strictly speaking, cannot be proven, just as incomparably simpler mathematical truths cannot be proven, in keeping with Gödel's Incompleteness Theorem. It can be shown, however, that the existence of the Maker not only does not contradict scientifically observable facts, but can also be logically derived from those facts with a very high degree of probability.

I call this position "technotheism", not in the sense that technology should be deified, but in the opposite sense: the rise of technology and the growth of its creative power makes more convincing the argument that our world itself has been created. Technology is a new theology, the 21st century's creative revelation. The fact that we shall be able to become designers of life and reason (which is where contemporary technology is gradually taking us, albeit without any guarantee of success), will most directly point to the existence of the Designer, although traditional faith does not really need any proof of that kind.

Technology is usually considered to be a sphere with the most radically atheistic outlook on the world. Indeed, if humans are able to rearrange the universe with their reason and energy, where does that leave the Maker? Why does He do nothing? How is His will manifested? Human activity, ever increasing throughout history, seems to leave less and less space for the Maker's activities. The Tower of Babel is being rebuilt, people keep climbing towards the sky, and, it would appear, there is no force which could bring down that tower of scientific technical progress, except for those forces which are unforeseen and therefore impossible to manage, such as natural catastrophes.

But let us ask: "Why should the success of technology disprove the existence of the Supreme Mind?" Why, on the contrary, should it not render even more realistic the possibility of such an all-powerful mind, which earlier seemed completely unthinkable to people who possessed only primitive tools? How can one, for example, explain to a farmer or a lumberjack that God can read all human thoughts? Or that a person, having died and turned to dust, can still outlive one's body and preserve the entirety of one's personality, the immortality of the soul? In days of old, technology was material, e.g. the axe, the plough, the hammer, and the sickle. In fact, the intellectual technology was only developed in the life of our generation, with the invention of computers, the electronic network, and simulated worlds. I personally found it easier to believe in the supernatural Mind after becoming acquainted with the possibilities of the artificial intellect (even though these still remain rather primitive, represented by computers and early corputers). If we are able to create something that resembles ourselves to such an extent, does that not increase the likelihood that we were ourselves created?

Unlike our ancestors, we understand how information can be gathered on a multitude of people within a small electronic device, and how our thoughts and habits can be calculated and predicted by powerful servers that accumulate information. For example, I type a word, and the computer knows in advance, even better than I do, which word I intend to type, on the basis of the frequency of the words I have used before. Or, when I use a search on Amazon.com, it offers me various items for sale that are connected, in a highly associative manner, with something that I searched for several months ago. The computer memorizes what I have forgotten, it knows what I want, it suggests what I can or must do, and it becomes a collocutor of my mind. For instance, recently I searched for a book by a certain author, and today Amazon offers me a different book by another author on a related issue. The whole worldwide network (or noosphere, or infosphere, or the world electronic databank) encompasses my intellectual demands and habits with increasing thoroughness, sending them back to me with some comments, offers, and associations, which become an active part of my mind, filling in the holes in my knowledge, memory, and, to a certain extent, my imagination. Given time, the network will find out about my sensory characteristics and habits, including my favorite smells and tastes. I will communicate with the network by using my voice, touch and gestures, which will also become part of the infinitely growing and, in its own way, creative memory of syntellect – the integrated intellect of people and machines.

Now, based on the experience of communicating with the latest technology, it is much easier for us to imagine how the Creative Mind can communicate with every living entity, read human thoughts and respond to them. To a ploughman who saw only the direct influence of one material object on another, it would be incomparably more difficult to imagine that whatever is secretly happening in his soul can become public and that "even the very hairs of [his] head are all numbered" (Matthew 10:30). How is it possible to number all the hairs of so many people? And how can one fathom all their thoughts? Where can one find such an all-seeing and omniscient spirit? How can it be everywhere and in everyone? Of course, a ancient peasant could simply take someone's word for this, without any explanation or proof, but for my contemporaries, such a notion of an all-seeing and omnipotent mind is no longer a matter of faith; it is the subject of an entirely reasonable, probable and well-founded assumption (again, I avoid the word *proof*). We now know how compact the means of storing information are, how a grain of matter can accommodate not only a plan of the future tree, but also – if the matter in question is electronic, e.g. a computer chip, a quantum system, and the like – thousands and millions of books, city plans, or information about all people, states, planets, etc. All information about the Universe and its every particle can potentially be stored in an electronic grain the size of a mustard seed.

Even such an erudite scientist as Richard Dawkins refutes the existence of God on the grounds that this hypothesis would entail an all-too-complex, all-encompassing mind:

> "A God capable of continuously monitoring and controlling the individual status of every particle in the universe *cannot* be simple. His existence is going to need a mammoth explanation in its own right. Worse (from the point of view of simplicity), other corners of God's giant consciousness are simultaneously preoccupied with the doings and emotions and prayers of every single human being – and whatever intelligent aliens there might be on other planets in this and 100 billion other galaxies." (Dawkins 2008, 178).

It is difficult for Dawkins to believe in such supernatural ability. But why does he not pay attention to the computer on his desk, which can find in a second what thousands of people, many of whom lived thousands of years before us, thought about any subject whatsoever. All it takes is to type a word and press a key. The Internet as we know it was created only two decades ago. It is easy to assume that, in the Creator's realm, during the course of the 13-14 billion years of the existence of our Universe (not to mention the unknown eternity before its coming into being), there could have appeared machines more perfect than the desk computer. Besides, God does not really need to regulate "the state of each and every individual particle of the universe": there are precise physical laws of the mutual influence of particles to take care of that. In a good laboratory, it is not necessary for a technician to constantly monitor all the details of a research process and regulate them manually.

If we can get information about the living and the dead in a mere moment by logging on to the Internet, what is strange about the fact that the Supreme Mind can hold inside itself the designs of not just our universe, but also the countless myriads of other universes, and can penetrate the greatest secrets of every person, the past and future of all sentient beings?

Primitive people could not understand how someone, who had died and vanished corporeally, could live on inside an invisible, intangible substance called "the soul". One could only believe in that by relying on speculations and promises that the soul would reach other worlds and find its place in heaven or hell. But to us, the coevals of CD-ROMs and electronic networks, it is much easier to comprehend rationally the difference between information and its material carrier. In the blink of an eye, information can be recorded from one disk on to another, from an old memory into a new one, or transmitted by cables or wireless signals. Does this not make the belief in the immortality of the soul, or (to use the language of the electronic era) the indestructibility of the informational matrix of a person, perfectly legitimate based on the data supplied by information science?

In ancient times, when the physical world was so domineering and unfathomable, it was hard to believe in the omnipotence of words that in some mysterious way determine the color of one's eyes and hair, hereditary diseases, and temperament. But does not contemporary genetics, which discovered the laws of heredity guided by the language of genes, confirm, in its own way, the fact that "in the beginning was the Word", i.e. that informational patterns precede corporeal existence and determine its properties?

For more than a century now, we have known of invisible rays that can be transmitted over infinite distances; we have known of the speed of light, and, more recently, of the dark matter and dark energy making up the greatest part of the Universe. And what of those mysterious black holes, possibly leading to parallel worlds? And what of the vacuum in which virtual particles are begotten? Or the Big Bang that led to the creation of our universe? And what of the astonishing balance of all the physical parameters of this universe, right down to their billionth parts, making possible our existence in it as sentient beings? Why should not science, relying on these physically verifiable facts, find a common language with theology?

Cognitive Faith

In the past, people believed in the religious image of the world in spite of its oddness and unreality. Why now, when it is becoming increasingly plausible, should we believe in it less? Is it only because we know more? In no way does this knowledge contradict faith: rather, knowledge absorbs and clarifies it. As a result, the vision of the omnipotent Engineer, the creative Word, and the immortal soul, which could earlier only be a matter of faith and superstition, now becomes incomparably more plausible. The religious development of humankind does not move from faith to disbelief; it moves from faith to knowledge. The time has come to speak of the *religiosity of knowledge*, not only of the religiosity of faith. The religion of knowledge is not a religion that bows to knowledge, but a religion that finds out from science, with increasing verifiability, about the things that the religion of old could only take on trust. I would say that the time has now come for *cognitive religion*, where cognitivism will play the same role that once was played by fideism, which holds faith superior to reason in discovering truth. Science and technology will not be the enemies of cognitive religion; they will not even be indifferent to it as an ostensibly "different culture" that has nothing to do with religion. Instead, science and technology will form a synthesis with religion since reason is increasingly in agreement with faith.

If reason abolishes faith at all, it is only to the extent of absorbing its content, becoming the believing reason. The scientific thesis, which holds that the Big Bang led to the creation of the universe *ex nihilo* is the object of not only physical, but also religious knowledge. The anthropic principle, which confirms that the universe was created so that humans could live in it, is religious knowledge as well. Separating information from its known material carriers, and allowing for an infinite diversity of these carriers, which transmit information about humans by extra-biological means, is a thesis of religious knowledge. Another thesis of religious knowledge is the idea that intelligence could in principle be implemented not only in biological neurons, but also in artificial devices, such as computational processors. One could go on for some considerable time enumerating all the ways in which religious faith enters the realm of contemporary science and turns into knowledge, at least approximately. That which people believed in the old days, we can now almost know, according to the words of St. Paul about how guessing will turn into knowledge: "For now we see through a glass, darkly; but then face to face" (1 Cor. 13.12).

Earlier, in times of the hammer and the hoe, the only thing left for people to do was to believe in the supernatural as a fairy tale, a marvel, or a myth. Technology brings the supernatural closer to us, makes it more natural to reason and thus rationally explicable. What lies in-between is no longer an abyss that can only be overcome by a leap of faith, but a high mountain that reason may gradually climb (even though it might never reach the top). We are apprentices who can, for the first time, assess the techniques of the Master's work – not to penetrate its secrets, but at least to understand where to look for them. In this respect, the history of science and technology constitutes a preparatory workshop where we gradually master the craft of engineering new worlds.

One widely discussed recent hypothesis on the artificial nature of our own world belongs to Nick Bostrom, a philosopher and the Director of the Future of

Humanity Institute at the University of Oxford. He finds it plausible to think that we are living in a computer simulation:

> You are almost certainly living in a computer simulation that was created by some advanced civilization. What Copernicus and Darwin and latter-day scientists have been discovering are the laws and workings of the simulated reality. These laws might or might not be identical to those operating at the more fundamental level of reality where the computer that is running our simulation exists (which, of course, may itself be a simulation). In a way, our place in the world would be even humbler than we thought. (Bostrom 2006, 38-9)[3]

According to Bostrom, it is highly probable that we exist in a virtual reality simulated by our technologically advanced descendants who, in a certain number of generations after us, will achieve the level of a superpower and superintelligence. However, this reasoning contains a fundamental flaw because the alleged civilization of the future that simulates us itself descends from us as simulations. There is a sort of circularity involved in this argument: we produce those posthumans who produce us. The proposition to be proved is assumed in one of its premises. If the world in which we produce our offspring is a simulation, then the source of this simulation cannot come from the same world, even in its future condition. To avoid this circularity, we have to admit the existence of another realm of being that is beyond our world of simulation, in the same way as the gamer and the gamer's computer belong to a different level of reality than the one simulated in the game. Bostrom himself acknowledges such a possibility, with all its theological implications:

> These simulators would have created our world, they would be able to monitor everything that happens here, and they would be able to intervene in ways that conflict with the simulated default laws of nature. Moreover, they would presumably be superintelligent (in order to be able to create such a simulation in the first place). An afterlife in a different simulation or at a different level of reality after death-in-the-simulation would be a real possibility. It is even conceivable that the simulators might reward or punish their simulated creatures based on how they behave. (The Simulation Argument FAQ – http://www.simulation-argument.com/faq.html)

As technology advances, humankind will find it increasingly difficult to manage without the notions of the Supreme Master of all computer games and simulations, designated as galaxies, planets, and the laws of nature in the language of the denizens of those simulations and the avatars of those games, in which the Author conferred upon everyone the gift of free will and the unpredictability of chance. As the virtual worlds that we create become increasingly authentic and lifelike, along with our own avatars within them, we will come to recognize more and more the features of such virtual reality around us and inside ourselves.

---

3   Bostrom, N. (2006): Do we live in a computer simulation? In: *New Scientist, 192 (2579)*, 38-39.

The Theological Paradox of Technical Advancement

If I can create an artificial mind or an artificial life that resembles me so very nearly, this increases the likelihood of my having also been created, and of the natural life and mind also being the products of artful engineering. This is not proof in the strict sense of the term, but rather a growing probability that the natural world as we know it, and we too, were created, just as virtual worlds are created by us and populated with our avatars.

We can recall Pascal's famous wager with its probability argument: If there is no God, then I, following the path of religion, deprive myself of a small amount of transient earthly goods. If, on the other hand, God does exist, I thereby obtain the infinite goods of immortality and heaven. This makes it more profitable for me to bet on God's existence.

In our case, it is not more profitable, but *more reasonable* to bet on God's existence, for the more the world manifests itself as our creation, the more likely it is that we were created ourselves. If we are capable of creating virtual worlds that are practically no different from the real one and possess the same sensual qualities, then what prevents us from positing that the physical world itself is a simulation? Insofar as only the artistic-conditional likenesses of the real world existed, for example, in the form of verbal descriptions and visual images, the difference between human hand-made creations and the "real" universe was evident. It was more reasonable then to assume that the universe had not been created, since it was difficult for the mind to imagine such power of creation. But if the ontology of our simulated worlds begins to approach the ontology of the "real" world in its complexity and sensory verisimilitude, then the creatability of this real world becomes increasingly probable.

It is more and more difficult to think of the world without the Maker – such is the conclusion of the entire technological evolution of humankind. Generally speaking, the more superior the mind becomes, the more able it is to recognize the superiority of another mind. Humility is not just a moral virtue, but an intellectual one as well. As human power to create an artificial mind and change the paths of evolution increases, we begin to come to terms with the idea of a power that created ourselves.

As a result, we will be compelled to recognize the evidence of the Engineer, the Designer, the Simulator, or the Gamer, i.e. the Somebody above us. This recognition may make spiritual and ritualistic forms difficult to imagine at present. This religious knowledge may be coupled harmoniously with traditional faith. Techno-theism may visit the temples of its ancestors in order to pray there, or it may turn them into museums. However, it cannot be questioned that science and technology possess enormous spiritual potential. Science discovers the laws of existence, while technology demonstrates the power of reason capable of creating a new existence on the basis of those laws. Based on these important testimonies of science and technology, it is difficult to resist a conclusion that the laws of existence were created by an even more powerful mind.

The Universe is much Bigger

I side with Carl Sagan, a prominent astronomer, author, and science populariser. Sagan was of the opinion that religion loses a lot by not accepting the achievements of contemporary science:

> How did it happen that in none of the popular religions did its followers, having taken a closer look at science, notice the following: 'Why, everything has turned out much better than we thought! The universe is much bigger than our prophets claimed – more magnificent, elegant, complex.' Instead, they repeat monotonously: 'No, no and no! Let my god be a small one – that's the kind that suits me.' If a religion – no matter whether it is old or new – praised the magnitude of the Universe that contemporary science has discovered, it would provoke rapture and command respect never even dreamed of by traditional cults. (1994, 52)[4]

But why does the biologist-atheist Richard Dawkins, who sympathetically quotes Sagan, react with his own 'No, no and no!' to each most sophisticated, magnificent, and non-dogmatic form of religion? Why do all those atheist naturalists repeat over and over again: 'Let my world be material only – that's the kind that suits me?' Why are they so unwilling to admit that, parallel with the visible matter, there also may exist a world that can never be adequately seen from the outside, but can only be experienced from the inside – the world of love, wisdom, sadness, conscience, repentance, desperation, and hope? Why do these materialists narrow their world far more than the most primitive believer, who still admits the existence of other worlds, miracles, and God's love and mercy? Why do they narrow their horizon down to natural selection and the "selfish gene" as the cause and bearer of all those aspirations, feats, and discoveries that make humans such fascinating, fantastic, creative, and self-sacrificing beings? Why do they address the question 'Why?' to believers only and not to themselves? What stops the propagators of scientific atheism from looking more closely at religion and seeing that science also loses a great deal by renouncing their possible joint action? Let me paraphrase Sagan's passage quoted above:

> Why, everything has turned out much better than we thought! The universe is much bigger, more diverse, and more spiritual than claimed by our materialist prophets, who recognized

---

4   Sagan, K. (1994): *Pale Blue Dot: A Vision of the Human Future in Space*. Random House, New York.

only matter given to us in our perceptions. This universe, the creation of the Supreme Creator who can address me personally and at the same time create myriads of worlds, who knows everything about me and loves me, who can do anything, but does not want to restrict my freedom, who has placed me to live in this world, but has revealed to me the paths leading to other worlds as well – this universe is incomparably more magnificent, elegant, and complex than can be imagined by any atheistic chemist or biologist, who allows only the existence of a scanty, tiniest part of a wondrously diverse cosmos.

Those who live in the twenty-first century should present such arguments to the learned opponents of religion whose atheistic views were shaped by the materialism and positivism of the nineteenth century. Let us patiently wait for a response, hoping that, in the century of technohumanism, a new mutual understanding of science and religion can be achieved. The more powerful humans become as the inventors of technology, the architects of the world, and the engineers of simulations, the more humble they find themselves in the presence of the Supreme Master.

Evolution and the Question of God and Morality
The Debate over Richard Dawkins

Dietmar Mieth

Presuppositions and Intellectual Discourse

In "The God Delusion" Richard Dawkins propounds and defends the thesis: "I am against religion because it teaches us to be satisfied with not understanding the world" (Dawkins 2008 a, b).

My position is the following antithesis: as a theologian I am not against agnosticism. However, some agnostics think that they comprehend the world adequately and have an explanation for all things and processes without exception, even for things and processes which they cannot sufficiently explain like the evolution of life. There are countless hypotheses about evolution as well as reconstructions of its initial developments and the conditions favorable to it – all of which should be examined – but there is no absolute certainty. In most of their explanations scientists do not take into account that advances in knowledge always entail an increase in nonknowledge as well. Advances in scientific knowledge open a vast area of potential directions in research. Surprises cannot be ruled out. The role of contingency cannot be underestimated.

Dawkins expressly rejects a dialogue on hermeneutics and scientific understanding with contemporary, well-known theologians. He is proud not to have read them (Dawkins 2008 b, 523), because they all base their reflections on the assumption that God exists. He is unwilling to comprehend or to accept that most of them suggest or even admit that this presupposition remains debatable. Moreover, he also asserts that these open-minded theologians do not represent the majority of believers – a strange argument on the level of intellectual discourse.

The medieval mystic Meister Eckhart (1260-1328) maintains that theology cannot be based on a proof that God exists, but that the theologian has the obligation to explicate his or her belief through "natural reasons". This means that one cannot remain faithful in the face of anything contradicting rational reflections without openness to the question of God in these scientific explanations.

The controversy over moral shortcomings and moral errors in the history of Christianity, for example, is not dependent on the question whether God exists.

Historical facts, ideologies, and their interpretations frequently are not dependent on the standpoint of a believer or nonbeliever. There are often concrete reasons related to belief or misconstructions of belief in a specific historical period which function as the background for reprehensible events and morally wrong actions. Dawkins' polemical characterization of Catholicism includes numerous passages on Hitler (Dawkins 2006, 272-278) as well as the reduction of Catholicism to the abuse of children and to restrictive educational methods (Dawkins 2006, 309-344).

One of the more interesting points in Dawkins' ethical presuppositions is the reversal of the authoritarian fallacy. The authoritarian fallacy is clear: an assertion is true or just, if the proper authority has proclaimed it. Then rational justification is unnecessary. In this context the infallibility of the Pope is often mentioned: admittedly, there are Roman Catholic positions which, in questions of truth and moral behavior (*fides et mores*), are not far from this kind of fallacy. But even the current, conservative Pope, Benedict XVI, appeals to the famous statement by John Henry Newman that conscience ultimately has precedence over the Pope. And conscience as an authority is not purported to be infallible. The only stipulation is that good reasons be reasons of conscience, reasons in which ultimate convictions are identifiable. The Protestant variant of the authoritarian fallacy is the strict, literal adherence to the words of the Bible, independent of their contextual foundation in historical and obsolete worldviews.

The "reversed" authoritarian fallacy posited by Dawkins means that an assertion is wrong if an authority proclaims it and asserts its truth. Dawkins' polemics against assertions made by religious authorities generally rest on the claim that an *authority* has proclaimed and urged adherence to a rule or ruling and then perhaps even persecuted the deviators. Although Dawkins is correct in inveighing against persecution and coercion, if the action undertaken by an authority is morally wrong, this by no means constitutes a proof that the original assertion is wrong. Anyone can do a number of things that are morally wrong, when acting in accordance with a truth of which he or she is convinced. But the morally wrong action is not sufficient proof that the underlying assertion is wrong as such. Nor is an assertion wrong if the reasons for its truthfulness have not been stated but have been replaced by pressure. It is clearly correct to point out this deficiency and to insist that reasons be given and submitted to rigorous intellectual examination. The difference between the lack of rational justification of an assertion, on the one hand, and the wrongness of an assertion, on the other hand, must be respected, even if someone is unable to rationally justify his or her assertions or if an authority replaces arguments with oppression.

Dawkins describes the ethics of the Old Testament as evil, appalling, nasty, and weird, the ethics of the Christian New Testament, on the contrary, as naive and ineffective (Dawkins 2006, 237-253). He rejects the historical and critical view of this history of theonomic ethics (Old Testament) and the common secular

Evolution and the Question of God and Morality 131

heritage of the Ten Commandments and human rights, for example. But does historical "evolution" encompass not only change but continuity as well?

In my opinion, a critical approach which merits the designation of scientific or scholarly discourse presupposes that the best counterarguments have been taken into account. Those persons to be criticized must be invited to discern whether their positions have been described accurately and fairly. If this is not the case, if the critical approach is reduced to the intended judgment, namely, right or wrong – then any and all arguments which seem to support my assertion are welcome.

A fair discourse is conceivable and possible. I would like to give two prominent examples: Erich Fromm, a famous representative of agnostic humanism, and Ernst Bloch, a Marxist philosopher, who explicitly recognizes the religious heritage of humanist thought. Dawkins mentions the argumentation of such respected agnostics in the afterword to the paperback edition of "The God Delusion" published in 2007, but his only counterargument is that humanistically-oriented religious thinkers are in the minority. Here he confuses questions of the majority of religious believers with questions of the influence of religious fanactics and religious power extolling fanaticism.

A Theory of Evolution or an Evolutionary Worldview?

Charles Darwin's theoretical explanation of the beginnings of life on this earth is not the same as an evolutionary worldview. A comprehensive evolutionary worldview requires the transformation of a theory into a an quasi-religious authority, which insists on its own exclusive truth in explaining all occurrences and developments in the history of humankind. In this manner history and the reflection on respectively the depiction of history – a very brief time in the course of evolution – serve as evidence for the evolutionary theory. But history is actually the cultural period of time in which humans reflect on their own actions in a microworld, or more specifically – before globalization – in a composite of different microworlds. In this historical microworld a macroperspective like a theory of evolution is not very expedient. The autonomy of historical development cannot be submitted to a theory of evolution even if there are potential analogies. However, in philosophical discourse, the term "analogy" actually means that the difference remains greater than the similarity. The time of evolution is a different time than the time of history. Potentially, in the time of history, if we can alter evolution through the implementation of our accelerated scientific and technological advances, history in the future may conceivably be another time of history.

For Richard Dawkins the evolutionary worldview becomes a kind of "not visible hand", a conception deriving from Adam Smith and his successors in the economic "Wealth of Nations". This "invisible hand" is the internal regulator of all that happens in life and in history: the egoism of the genes and selection are examples of paradigms, which allow us to not only understand the world and human destiny, but which regulate our moral insights as well. This conception characterizing the market economy has a parallel influence on moral liberalism.

In his relativizing critique of Christian ethics, Dawkins apparently has other authorities on his side: Bismarck, for example, said that the Sermon of the Mount from the New Testament does not explain how to structure politics or political action. This is true if politics is nothing more than the struggle for power. Max Weber has spoken of the so-called *Gesinnungsmoral*, the ethic of conviction, in which the moral intention always overrules the questions of effect and effectiveness. However, Max Weber did not envision a morality with the sole intention of weighing risks and benefits without principles as criteria. This popular misunderstanding of his approach is often quoted by journalists.

Where Dawkins' Provocation is Justified and What is to be Learned from Religion

It is evident that all religions declaring God's love for humankind or advocating belief without violence have nevertheless supported hate and violence in the realization of their own projects in different historical epochs. Religions often were not able to assert and justify the truth which they wished to promote in moral practice through nonviolence. Military conquest, the Crusades and other holy wars, struggles against heresies, impediments to the rise of human rights: the history of religions is full of inconsistencies in the realization of religious truth and the realization of the vision of the perfect life and moral social institutions.

However, the realization of religious visions of the perfect life and of moral institutions has often been historically successful. The concrete realizations of religious visions are often models worthy of emulation, although they occasionally entail indications of abandoned humanist implications of the asserted truth. Religions are more ambiguous than they should be. This may be inherent in the human condition: in the name of the perfect, the realization of the imperfect is a grave danger. (Robespierre's Reign of Terror is a well-known example, though not in the area of religion!) The conclusion is that religion needs to participate in a open discourse monitored by moral reasoning and moral experience. The tradition of "natural law" in the Scholastic schools of Paris in the Middle Ages and the autonomy of reason propounded by Kant are examples of such counterbalances, which theology in different religions must continue to employ. Secular moral reasoning can also be ambiguous. Therefore, a good working relationship and mutual critical interaction between moral philosophy and moral theology are absolutely imperative.

One of the best theological arguments for this mutual process of critical questioning and monitoring is to be found in Meister Eckhart's vernacular sermon number 86, in which he asserts that the pagan philosophers (here he is thinking of Aristotle) are more precise in their concrete ethical reflections than the theologians. Meister Eckhart attributes this to the fact that their nontheological viewpoint (which he calls the "natural light of cognition") clearly reveals the distinctions which must be made in moral reasoning. Intriguingly, religious approaches to reality guided by the "supernatural illumination" of cognition cannot see these necessary distinctions clearly, because they are blinded by this illumination and cannot distinguish between light and shadow, right and wrong, in questions related to the "virtues".

The Question of God, the Moral Tradition, and the Future of Ethics

Ethical rules have their roots in social and juridical rules. In ancient civilizations, for example, in Mesopotamia, these rules were combined with religious rules by a so-called covenant between state authorities and religious authorities. This covenant was ascribed to God's initiative and accepted on the basis of the belief in God's responsibility for all that happens on earth. This dimension of God was not questioned for a long time, even when a plurality of gods were instated and respected. In this historical context the biblical development of rules is apparently not unique. However, in the unbroken tradition of the Bible, there is a continuity based on early, common rules like the Ten Commandments, the vision of justice for the oppressed in the prophecies, the moral wisdom in resolving the problems of daily life, the experience of being a moral self with a moral identity also providing a social identity.

In ancient history the question remains: is God the composite of all understanding of what is good ? Or is God more complex? „*Er [Gott]war nicht das Gute, sondern das Ganze*" ("He [God] was not the Good, but the Whole"), as the German author Thomas Mann writes in his biblical tetralogy on Joseph. This implies not only the distinction between the good and the right, but also the belief that God cannot be totally comprehended through moral categories. On the one hand, the religious answers to the question of what constitutes a meaningful life were not always morally practicable or tenable; on the other hand, the question of what is good and right does not encompass the question of hope central to religious belief. Belief is hope for eternal love, and is not a specific moral instruction. Belief in God cannot constitute a final foundation of ethical theory and moral practice, and moral insights do not guarantee certainty about God's existence.

Is "God" indisputably a figure of competence in the moral tradition but, following his demotion to a "delusion", worthless for ethical discourse in the future? Ethics often reflects on processes important for the future. In bioethics the question of enhancement is often a major issue. Questions related to financial crises, corruption, doping are also daily concerns of ethics. Sustainability and globalization, human dignity and human rights, the process of justice and peace, migration ... the topics related to our future are almost endless. What role will religion play in the reflection on these processes influencing the future?

One point to consider in this process is the conception of a "global ethic", formulated by the Catholic theologian Hans Küng in Tübingen. Another is the process called "peace, justice, and the care of creation". Both conceptions offer special programs for interreligious education and are open to humanist and agnostic approaches to ethics. What is problematic about Dawkins' argumentation is his insistence on exclusion. Exclusion is not always wrong. When Dawkins chooses to exclude certain proofs of God's existence, which "worship" the "gaps" of scientific explanations, because they seemingly offer God a place to exist, I can agree with him (Dawkins 2006, 125-134). However, in his entire chapter on arguments for God's existence, he has forgotten to mention one crucial argument: that many people believe in God and feel that this religious belief enables them to lead a good human life. This is clearly not an argument which can settle the debate over God's existence. There are also people – for example, in the postcommunist world – who assert that life without God is sufficient, even for a good moral life. Different life experiences and positions cannot determine what is true or false, but they can coexist in an atmosphere of mutual respect, despite these differences.

I will assert that an ethics of autonomy is necessary to criticize religious approaches to ethics. However, I am exceedingly aware that the concept of an ethics of autonomy or a secular (nonreligious) ethics does not adequately deal with a number of issues, issues, which, in my opinion, are crucial for the future.

Philosophical ethics does not generally respect the phenomenon of finiteness (finitude). There the talk of "contingency" is not talk of the same thing, and often means "coincidentality", events which happen by chance and by accident. The anthropology of finiteness has been almost totally forgotten. If we want to optimize the human being, for example, it is inevitable that what we want to optimize, that is, the old, will actually determine what will be the new.

Also forgotten ist the question of personal guilt. It is usually relegated to psychologists and psychoanalysts. The question of collective guilt was the last ques-

Evolution and the Question of God and Morality 135

tion related to guilt that was contemplated by certain philosophical schools. Forgiveness is most frequently relegated to theologians. Only a few philosophers, who are also physicians, discuss suffering.

Theological ethics is not justified by these "gaps" in philosophy. Its role in philosophical discourse is to direct attention to topics that are often dismissed or ignored, topics which are undeniably questions of the future as well. (Here German theologians often quote Jürgen Habermas and his reconstruction of the role of belief; however, Habermas argues as a philosopher who misses the religious content but insists that it be monitored by philosophical discourse.)

In the second part of my reflections I have chosen the question of prejudice as an example. Although this question is often broached with a polemic against religious faith like Dawkins' polemic, there are ample instances of prejudice without an underlying religious belief. The field of the biosciences and bioethics, in which I worked for more than twenty-five years, will serve as the context for my reflections.

Science and belief – a dialectical question: Preliminary thoughts on prejudice in the biosciences

Max Planck wrote (^{7}1969, 248):

"If, therefore, in [...] many [...] cases belief proves to be that power which first leads the accumulated individual scientific data to real effectiveness, one may go even further and claim that beforehand, in the process of collecting the data, a foresighted and inquiring belief in the deeper interrelationships can actually be quite helpful. It shows the way to proceed and sharpens the senses. For a [...] researcher who follows his experimental protocol in the laboratory and carefully examines the accumulated data, the progress of the work [...] is facilitated in many cases by a certain, more or less clearly conscious, special intellectual conception, with which he designs his experiments and considers and interprets the obtained findings."

That modern science–according to its own self-understanding–is without religious presuppositions does not mean that it does not require a preceding fantasy or an option formulated as extensively as a "belief", as Max Planck has noted. Admittedly, this has critical limitations and must be realized through reflection. Above all, the nonreligious foundation of science is to be distinguished from an unnecessarily assumed criticism of religion. The literary scholar Peter von Matt has summarized this distinction concisely:

"There, too, where it seems the clearest, in the area of the strict sciences, for which there absolutely may not be theological implications, the departure from a theological, salvation-oriented worldview originally was not by any means carried out as an act of unbelief but as a purely scientific rule of procedure. This rule said that nature was to be investigated under disregard of the question of a Creator. This means science is to be conducted *as if* he were not to exist – 'etsi Deus non daretur' – not, however, 'quia non datur', *because* he does not exist." (von Matt 2006, 245)

This distinction is to be kept in mind when quarrels arise between science and worldviews. It is possible that this dispute, seeing as to how it is continually restaged, repeatedly finds enough motives on both sides. In my opinion, it is unnecessary. It is wiser to keep science and belief distinct from one another. This, of course, means that when premises take on the dimensions of belief, they must be examined critically. It does not mean that presuppositions about belief, which do not affect the independence of the scientific method, must be combatted.

Prejudices Within the Biosciences: Their Expression in Language and Social Consequences

Through several examples from the biosciences I will demonstrate how such prejudices enter the collective consciousness and, beyond this, can affect society. Here I am proceeding from the observation that offensive scientific policy-making prefers instruments of language politics, which explains why scientific prejudices often take on specific linguistic forms. How greatly negative prejudice attempts to influence the language of negotiations on scientific progress can be demonstrated on the example of "reproductive" cloning since approximately 1997. The ban on "reproductive" cloning is even included in the Charter of Fundamental Rights of the European Union. The background is as follows: This specific use of the expression "reproductive" first arose in 1997, as all the commissions were discussing "Dolly", the cloned sheep. It entailed a shift in semantics. Until 1997, it was established that ovum and sperm cell each belong to the so-called "reproductive", biological "substances". The word "reproductive" thus began with egg cell and sperm cell, just as it is still included in the expression "reproductive medicine" today, which, of course, constitutes a fundamental technology in biotechnology. When we speak of reproductive medicine or "procreative" medicine (*Fortpflanzungsmedizin*), we mean the assisted fusion of egg and sperm cell in vitro. Thus, egg and sperm cell are reproductive. Behind the expression "reproductive' cloning", however, is the implicit assumption that ovum and sperm cell are no longer understood as being reproductive. "Reproductive" cloning signifies only the carrying to term after implantation in the uterus; a cloned embryo is carried to term following implantation. A ban on "reproductive" cloning, as specified in the UNESCO Universal Declaration on the Human Genome and Human Rights (1998), does not, then, encompass the in vitro area.

When reproductive cloning is prohibited, we have to read the subtext as well: the so-called "nonreproductive" cloning in vitro is not included. "Nonreproductive" cloning is the cloning of embryos without implantation, for example, for purposes of destructive embryo experimentation. When the Charter of Fundamental Rights of the European Union forbids reproductive cloning, the question of the right to life of embryos is left open at the same time in a subtext.

Negative structures of prejudice are also encompassed in the expression "'therapeutic' cloning". In scientific communication in biomedicine the term *therapy* is widely used, although no therapy exists to date. "Therapeutic" cloning is cloning with a specific research intent, possibly with later consequences for therapeutic models. However, the actual steps leading to this goal are still unclear. One does not say what one knows but what one wants to know, in a way as if one already knows this. Cloning itself is not a therapy but a research procedure. The expression "'therapeutic' cloning" is thus a misrepresentation entailing a high degree of dishonesty. This is tolerated in the hopes of finding broader acceptance. Frequently, increasing acceptance is in turn–as in the present case – accompanied by increasing public pressure. A society in which, for example, the conviction, nourished by language politics, has gained acceptance that "therapeutic" cloning will with absolute certainty lead to a cure for diseases to date incurable, will do everything in its power to follow this path. That this path possibly could be unsuccessful is considered just as infrequently as the possibility that alternative strategies could eventually be equally promising and even less ethically questionable. This example illustrates how bioscientific prejudices can have adverse effects on society by blocking awareness of potentially better alternatives. (To avert this danger, the U.S. President's Council for Bioethics, for example, despite extensive differences in the ethical assessment, rejected the expression "'therapeutic' cloning" in a position paper in 2002 ["Human Cloning and Human Dignity: An Ethical Inquiry"] and replaced it with "cloning for biomedical research.")

Distinctions in language are often just as significant as real images. This is well illustrated in the differentiation between in vitro and in vivo. One distinguishes "in vitro", i.e., in the test tube or in the petri dish, where the embryo is exposed, from in vivo, which means "in life". This is not to imply that in vitro is not a living being. But in the choice of words, in language which first explicitly names life outside of the test tube, one reinforces the prejudice that laboratory research merely involves biological material and not "real" life. Another example: In the European directive on patenting biotechnological inventions, mention is consistently made of "biological material", although it is dealing with "life". Animals and human genes, too, are part of this life, and, consequently, are not biological "material". The total materialization of the concept of life is becoming widespread. For many this is only a "methodological" materialization, because they are otherwise incapable of recognizing and describing the relationships between cause and effect. If one asks what definition constitutes the foundation of the life sciences, the answer is: research on living systems or organisms. No probing questions are asked: for example, how do organisms differ? As long as this specific theoretical language of the life sciences is methodological, as long as it recognizes its own relativity within the context of the various disciplines and approaches, it is not problematic. But if this language becomes a dominant paradigm for the language about "life" and if we derive our comprehension of life in general from this, we have every reason to reflect seriously on this tendency.

Here, too, is ultimately the level where one encounters talk of the embryo as a "mere heap of cells", a representation not uncommon in the biosciences. If one has a "transcendental" conception of the individuality of the human being, which precedes every assessment and every empirical qualification, then the question whether one can see some aspect of this in the embryo is secondary. Decisive are instead arguments of the potentiality, continuity, and identity of the single human life. People ask, for example: What does the embryo, which I developed from, have to do with me? A disabled person, to whom one says that one would like to choose an embryo because he or she has the same disease or impairment, replies: "Then you would have chosen 'me'". The question whether the disabled are discriminated against through such a selection in the test tube is controversial and calls for a separate discussion. (Cf. van den Daele 2002) The continuity evoked in the answer of the disabled person, on the other hand, is recognized everywhere purely intuitively. Questioned about immune defense in the context of the implantation of stem cells derived from a human being's own body material, a stem cell researcher participating in one of the many ethically controversial debates on in vitro cloning told me: "Naturally, the immune defense of the human being is activated, it's a foreign individual, of course." This is a spontaneous answer without much thought, and is precisely for this reason illuminating. What is meant are the genetic resources in the so-called "mitochondria", inside a foreign egg cell membrane, which one also needs in cloning technology, although this strives for the replication of a person's own cells for the regeneration of organs.

If every human living thing as "human being" has the right to human dignity, ethicists can still argue over gradualistic positions, diminishing the chances of a totally convincing argument here, but it is possible to rule out the designation of an embryo capable of such a high degree of differentiation as a "mere heap of cells." Nor is this correct under premises that want to disallow every metaempirical definition of the human being. When an embryologist answers the question as to when human life actually begins with the flippant remark: "When I can see the nose", then he or she has reduced the biological phenomena to an external im-

Evolution and the Question of God and Morality 139

pression. This is consistent with an extensive prescientific tradition. (Cf. Willam 2007; Rager 2006; Hilpert/Mieth 2006).

Responsibility for Progress in the Sciences: What does Progress actually signify? What constitutes a Scientific Advance? Accountability for Progress in every Scientific Advance

To speak of single scientific advances, of progressive steps (*Fort – Schritte*) is a subtle suggestion to seriously consider once more what progress (Fort*schritt*) actually means. From the perspective that essentially everything new is better than the old, progress has more or less become a global model which has deeply ingrained itself in our society as a "progress mentality." Much is packed into this model, for example, the alliance of modern society with science, technology, and economy. It almost seems as if this alliance is irrevocable, since it is the precondition for progress so comprehended and the means that produces this progress. But does progress really exist in this generalizing singularity? Or is progress as a general idea a prejudice? Are there only single scientific advances toward progress? Progress as a whole is conceivably a promise of the modern age, during the course of which technological progress has always been considered social and cultural progress at the same time. There is, however, adequate opportunity for reexamining this promise retrospectively, namely, so that single scientific advances are individually scrutinized on the basis of their effects on society as as whole.

What single scientific advance is truly worthy of being called a step toward progress? How can progress be measured? Are the valid criteria here the advances in knowledge within the previously established framework of a scientific paradigm? Or are the valid criteria those of technological feasibility or those of economic viability? Or perhaps the criteria of compatibility with human, social, intellectual achievements? This topic has already become universally significant in the context of the central concept of sustainability. Thus, it is a good therapy for the prejudice associated with general conceptions of progress to unwrap the "progress" package and not to ask whether it is a matter of progress in the sense of a general trust or even belief in progress, but how the assessment of proposed or already initiated scientific advances might look in the individual case. The transition from single scientific advances to problematic generalizations is always to be examined individually.

Truthfulness and Transparency instead of the Normative Power of the Fictive

Whoever works with scientific exactitude does not deny the uncertainty of the attainability of options in insight and in application, the simultaneous ongoing awareness of the unknown ensuing from more exact knowledge, and the long stretch ahead to the attainment of this uncertain goal. The great degree to which this is overlooked in the momentum of a new paradigm became clear to me in an exemplary incident related to "gene therapy" as early as 1989, when I had the opportunity as a member of a group of representatives from Catholic universities to speak with the renowned hematologist French Anderson about experiments in "gene therapy" at the National Institutes of Health (NIH) in Washington, DC. Immediately striking was the discrepancy between the posters mentioning therapies for AIDS, cancer, and immunodeficiency, and the laboratory experiments centering on mouse blood. Consequently, I attempted – as a layperson – to imagine a scenario, which establishes a scale from 1 to 100, the number 100 designating the actual attainment of a therapy. At that time, I envisioned, with regard to so-called gene therapy – similar to today, with regard to stem cell research – colossal adventures awaiting science. My question then was to what extent and with what degree of acccuracy, given that laboratory experiments at that time attained 1 to 5 on this scale, a prediction for the achievability of 100 could be made. The answer was: "I don't know, but I do believe it's possible." Here a curious contradiction is evident: between the presuppositions of belief which science criticizes and the presuppositions of belief which it considers imperative for progress in knowledge. A criticalness, which does not put its own house in order but points to the prejudices (prejudgments) of others, ultimately cannot be convincing.

It is obvious that a potential development of research cloning on human beings also entails great reservations: in the speculative application of animal experimentation on human beings we cannot predict that healthy cells, if we implant them in a diseased organ, can make this organ healthy again. The great hopes for therapeutic cell transplantation rest on shaky foundations, which collapse again and again. In a time period in which we are conducting initial experiments in the areas of animal cells or initial experiments in the area of human cells, are we capable of predicting that we will actually attain a "therapy"? And that there are no alternatives to this way? (Bentele 2007) Apparently not.

Freedom and Responsibility in Science: The Problem of Prejudice

Freedom of research as well as the freedom of the individual researcher from external influence must be guaranteed. Freedom does not relieve one of responsibility but *is* an essential precondition for its acceptance. This is why the question is justified how free research is or can be at all, if it is dependent on astronomically expensive apparatus and equipment as well as on the choice of a research topic currently being funded. One cannot compare an abstract conception of the freedom of research with concrete conceptions of ethical obligations in research. Instead, the obligations must be critically scrutinized in many respects in order to examine which obligation of this kind contradicts which obligation of that kind. The objection will be raised that with technical and financial limitations in research, there is more choice than in the adherence to ethical restrictions, which can be imposed by law. This choice, too, is to be analyzed for dependencies. And the power of emotional and quasi-metaphysical research expectations also constitutes a considerable limitation.

Responsibility in research possesses an internal and an external dimension. Internal scientific responsibility means that rules of independence, fairness, and scientific integrity are observed and monitored by the scientific community via internal rules. There, the freedom of conscience of the individual, and that of the dependent researcher as well, must continue to be ensured. External scientific responsibility means that the modes and consequences of research are ethically tenable for all directly and indirectly affected persons and are open to examination by society. However, external responsibility can hardly be exercised alone within the research discipline. Even in transdisciplinary research, there is often a lack of auxiliary competence, which is why research programs in the EU member countries and in other countries are often announced with supplementary ethical, legal and social expertise (the ELSA concept: ethical, legal and social aspects). This requires the combination of highly professional and, at the same time, ethically reflective competence. Augmentation of this competence through the broad experience of religion with its own prejudicial structures – recommended by an EU conference on bioethics in Amsterdam at the beginning of 2005 – seems thoroughly appropriate in light of our considerations (Cf. Mieth 2006).

Conclusion: Overcoming Prejudice as the Lack of Self-examination and as the Incapacity for Correction

Prejudice in a negative sense means prereflectivity and the incapacity for correction. Therapy cannot be found in radically rejecting prejudices and aggressions as such but only in analyzing the basic anthropological patterns and in consciously revoking the structures of decline in prejudice, which, incapable of self-correction, practice constant self-affirmation. Thus, therapy can only begin with the recognition and acknowledgment of our prejudicial structure, on the one hand, and, on the other hand, with the rejection and exposure of the self-affirming prejudice incapable of correction, impacting or at least lessening its effectiveness.

Modern science is associated with the effort to fight prejudices and to avoid their potential generation through pressure for methodologically transparent action. The models with which this effort was theoretically formulated are worthy of respect, even though the discourse on the verification and falsification of scientific insights will continue. The problem of prejudice, however, is unrelated to this monitored search for truth in science and is not resolved by it. It is, however, present in scientific practice. Accordance of scientific practice with the fundamental theoretical presuppositions to which one subscribes is one of the ethical stipulations for responsible science.

The nucleus of prejudicedness in scientific practice is based on the assertion that because science as such does not exist under the spell of prejudice, options and prognoses can also be proclaimed in its name, which tend to have either a selective and discriminatory character, by measuring the prejudicedness of others on their own imagined unprejudicedness, or a prophetic rather than a scientific character. Since, however, anticipation is integral to knowledge, it is imperative to not only examine the coherence with which insights in this framework arise, but also the contingency of one's own anticipatory paradigm. Science, which frees itself from the binding character of its conclusions and extrapolates these conclusions by means other than those of scientific inquiry, confuses the categories, above all, if it attempts to influence social conditions and social responsibility.

In his best-selling book "The God Delusion" Dawkins bases his polemical attack on certain well-known, morally reprehensible actions and moral catastrophes in the history of (Western) religion. He posits an "authoritarian fallacy": whoever speaks with authority on moral questions is always wrong, independent of his or her arguments! He makes no distinction between a lack of rational justification (which is to be criticized) and a morally wrong instruction. Ultimately, Dawkins epitomizes a prejudice purporting to be a scientific explanation, which is actually a presupposition.

Bibliography

Bentele, K. (2007): Ethische Aspekte der regenerativen Medizin am Beispiel von Morbus Parkinson. Reihe: Mensch – Ethik – Wissenschaft. LIT, Münster.
Dawkins, R. (2006): The God Delusion. Bantam Press, London.
Dawkins, R. (2007a): The God Delusion. paperback edition (Black Swan). Transworld Publishers, London.
Dawkins, R. (2007b): Der Gotteswahn. Ullstein, Berlin.
Enquete Commission of the German Parliament (2005): Recht und Ethik in der modernen Medizin. 6. September 2005: http://dipbt.bundestag.de/dip21/btd/14/090/1409020.pdf (14.02.2012, 3pm)
Gadamer, H.-G. (1972): Wahrheit und Methode. Mohr Siebeck, Tübingen.
Habermas, Jürgen (1971): Hermeneutik und Ideologiekritik. Suhrkamp Verlag, Frankfurt.
Hilpert, K./Mieth, D. (eds.) (2006): Kriterien biomedizinischer Ethik. Theologische Beiträge zum gesellschaftlichen Diskurs. Herder, Freiburg im Breisgau.
Kaufmann, B. (1965): Der Mensch im Banne des Vorurteils. Brockhaus, Wuppertal.
Krüger, O./Sariönder, R./Deschner, A. (eds.) (2003): Mythen der Kreativität. Das Schöpferische zwischen Innovation und Hybris. Lembeck, Frankfurt.
Mieth, D. (2006): The role and backgrounds of religious, ethical, legal, and social issues in the progress of science. In: Jeanrond, W.G./Mayes, A.D.H. (eds.): Recognising the margins. Developments in biblical and theological studies. Essays in honor of Seán Freyne. Columba Press, Dublin, 321-334.
Planck, M. (1969): Vorträge und Erinnerungen. Wissenschaftliche Buchgesellschaft, Darmstadt.
Rager, G. (2006): Die Person. Wege zu ihrem Verständnis. Academic Press Fribourg, Freiburg, Switzerland.
Rehmann-Sutter, C. /Düwell, M. /Mieth, D. (eds.) (2006): Bioethics in Cultural Contexts. Reflections on Methods and Finitude. Springer, Dordrecht.
Pelinka, A./Bischof, K. /Stögner, K. (eds.) (2009): Handbook of Prejudice. Cambria Press, Armherst, NY.
Van den Daele, W. (2002): Zeugung auf Probe. In: Die Zeit 41, 02.10.2002, p. 34.
Von Matt, P. (2006): Die Intrige. Theorie und Praxis der Hinterlist. Hanser, Munich.
Willam, M. (2007): Mensch von Anfang an? Eine historische Studie zum Lebensbeginn im Judentum, Christentum und Islam. Studien zur Theologischen Ethik 117. Herder, Freiburg, Switzerland and Freiburg im Breisgau.

Evolutionary Theory Applied to Institutions The Impact of Europeanization on Higher Education Policies

Vojin Rakic

The Problem and a Tentative Theoretical Model

What was the impact of European integration on higher education policies of its member states and the subsequent adoption of the Sorbonne Declaration (1998) and Bologna Declaration (1999)[1]? To answer this question I will focus on six states of the European Union: Germany, the Netherlands and Belgium/Flanders[2], Great Britain, Sweden and Finland. The selection of these six cases is based on the fact that they represent states that are culturally related, but that enclose both the founding members of the European Community, as well as "newcomers" to the European union of states. Hence, the higher education policies of the selected cases:

- were not isolated from each other;
- contained similarities that might not only be attributed to their membership in the European Community/Union, but possibly also to other mechanisms that resulted in a convergence of their higher education policies.

An attempt will be made to assess whether the higher education policies in these six countries were converging, diverging or not changing at all – before the signing of the Sorbonne and Bologna Declaration. After proposing an answer to this question, I will deal with the issue whether change (convergence or divergence) or its absence was a consequence of EU policies or of other circumstances. First, however, the concepts of convergence/divergence (homogeneity/diversity) in higher education will be discussed. Borrowing from evolutionary theory, I will attempt to prove that institutional imitation, rather than EU policies, was a primary mechanism responsible for observed convergence of higher education policies of

---

1 The expressions "European integration" and "Europeanization" I understand as coterminous. They refer to wider integration processes than those that are merely a direct result of European Union policies.
2 After the federalization of Belgium in 1989, higher education policies were almost completely transferred to the regional level and the level of the language communities. I will concentrate on the higher education system in Flanders, treating it as a part of the Belgian system before 1989.

European states[3]. In the last section (termed "epilogue"), the paper will briefly return from higher education policy issues to evolutionary theory in general, advancing the thesis that *adaptation mechanisms* are not incompatible with *creation mechanisms*: isomorphic processes in higher education policies do not exclude the existence of "policy creators", as adaptation mechanisms in nature are not incompatible with the idea of "intelligent design".

A body of literature in the field of higher education deals with convergence/divergence (homogeneity/diversity[4]) at the level of higher education *institutions* and at the level of *disciplines*. Similarly, Birnbaum (1983) distinguishes two forms of diversity: *external diversity* is based on differences *among* higher education institutions, whereas *internal diversity* denotes differences *within* higher education institutions. I will not concentrate on either, but on a third form of diversity, that is on differences among higher education *policies*. The term most suited for it might be *systemic diversity*.

Changes of one form of diversity can be accompanied by changes of another form of diversity *in the opposite direction*. For instance:

divergence among higher education institutions within a higher education system can be the result of converging policies at the systems level. As an example, EU policies that promote diversity can lead to convergence at the systems level (the EU member states adopt similar policies aimed at the promotion of diversity) which might result in divergence at the level of higher education institutions within the member states;

convergence among higher education institutions *within* a system (external convergence) might be the result of diverging policies at the systems level (systemic convergence). For example, different policies of a given number of governments can result in higher education institutions affected by the policies of those governments becoming less diversified.

It is therefore crucial:

not to confuse convergence or divergence among higher education institutions (or even within a higher education institution) with the same phenomena at the systems level, i.e. with higher education *policies*;

to understand that the observation of convergence or divergence is contingent upon the level of aggregation (e.g., divergence may take place at the level of higher education institutions, whereas convergence may be observed at the systems level).

To clarify the distinctions mentioned in the foregoing paragraphs, I will discuss three influential perspectives on the question of diversity/homogeneity and di-

---

3   The argumentation will to some extent be based on Rakic (2001) and Rakic (2002).
4   I will use the terms homogeneity/diversity to denote a particular (static) state of affairs, whereas convergence/divergence and homogenization/diversification (or differentiation) I will employ to denote a particular process, i.e. a dynamics. For an extensive review of diversity concepts, consult Huisman (1995).

Evolutionary Theory Applied to Institutions 147

vergence/convergence in higher education. The first, internal perspective, is promoted in a text by Burton Clark, the second, systemic perspective, is discussed by Guy Neave, and the final, mostly external perspective, can be understood nicely on the basis of a text by van Vught[5].

Burton Clark (1996) elaborates on the *internal perspective*, which emphasizes the basic unit in higher education institutions, i.e. the academic discipline. According to him, differentiation is a natural process in higher education ("the creation of finely distinguished subcultures in academia is a natural process" [ibid., 19])[6]. Clark writes even about "balkanized authority" in the higher education field (ibid., 20). In his own words:

> "higher education is a differentiating society *par excellence*. It adjusts internally to increasing arrays of input demands and output connections by greater specialization in its production units and the programs they offer. Adapting to the changing contours of rapidly expanding and highly specialized knowledge, it creates more varied types of academic tribes [...] Higher education is pre-eminently an internationally shaped component of modern and modernizing societies [...] The dynamic of differentiation is a powerful root cause of the tendency for higher education to be a self-guiding society. Governments and other patrons will increasingly find higher education to be a contentious area highly resistant to command and control" (ibid., 24).

Apparent convergence processes, Clark explains by his notion that what we interpret as convergence is in fact divergence along a continuum:

> "Academic drift is a converging form of drift; second and third sectors of institutions converge on a first sector as they seek to emulate its ways and to gain similar power and prestige [...] They only achieve 'weak emulation'. Individually the emulating institutions add to differentiation as they become variously sorted out along a continuum of degrees of difference" (ibid., 23).

Guy Neave (1996) concentrates on the systems level. He directs his attention to higher education *policies*, in particular to the national and supranational policy levels. The "European dimension" has to be taken into account seriously in research on higher education policies of its member states,

> "even if it is not entirely clear at present how it bears upon the issue of convergence versus divergence" (ibid., 29).

In others instances, Neave appears to believe that EU policies do result in convergence among national higher education policies:

---

5 My discussion of the three perspectives on the basis of texts by Clark, Neave and Van Vught, does not suggest that these authors are somehow proponents of the three perspectives. They only used the mentioned perspectives in their specific texts, without suggesting that any one of the three is the only right or most important one.
6 For a more elaborated statement of Clark's position, see Clark (1983).

"There can be little doubt of the influence the European Commission is already exerting at establishment level and, no less important, as a force for convergence between systems" (ibid., 31).

"The 'European dimension' – or what is termed as 'Community level' action in the inimitable jargon of Brussels – is explicitly constructed around the ultimate aim of economic, financial and industrial integration in which convergence stands as the highway leading to this ultimate goal. These three areas of activity cannot, unless one believes in the convenient legend of the university as an ivory tower, but move higher education systems in the member states towards a similar condition" (ibid., 38).

However, Neave also observes the difficulties in assessing if we deal with convergence or divergence:

"In effect, the perception we have of a particular system and *a forteriori* of whether it is diverging or converging is largely a function of where we focus our attention. Like the Cheshire Cat, sometimes we see a tail, sometimes a head and sometimes just the grin" (ibid., 28).

"homogeneity, convergence, diversity and variety are not absolutes. They are temporary and unstable conditions which always bear watching and, for that reason, will always be part of the warp and weft to our fields of enquiry" (ibid., 39).

Van Vught's (1996) explanatory framework is to a significant extent derived from three theoretical perspectives from organizational theory: the population ecology perspective, the resource dependency perspective and the institutional isomorphism perspective (ibid., 43). His emphasis is on the environments (primarily governments) in which higher education institutions operate, as well as on academic norms and values (i.e. the academic culture):

"The various empirical studies appear to underline the notions of the theoretical framework presented earlier. According to the authors of these studies, environmental pressures (especially governmental regulation) as well as the dominance of academic norms and values (especially academic conservatism) are the crucial factors that influence the processes of differentiation and dedifferentiation in higher education systems. In all cases, the empirical observations point in the direction of dedifferentiation and decreasing levels of diversity. The overall impression is that, in empirical reality, the combination of strict and uniform governmental policies and the predominance of academic norms and values leads to homogenization" (ibid., 56).

All three approaches insist on change. Thus, they attempt to establish if there is convergence or divergence in the field of higher education. However, the three approaches concentrate on different levels of analysis. Clark primarily focuses on academic disciplines, i.e. on *internal diversity*. Neave's interest in his chapter is mostly directed toward (national and supranational) policies, i.e. toward *systemic diversity*. Van Vught's chapter deals mostly with developments *within* higher education systems, i.e. with *external diversity*. Clark observes divergence, Neave and van Vught mostly convergence (though with reservations). Since these three approaches deal with different levels of analysis, however, they are not mutually

Evolutionary Theory Applied to Institutions   149

exclusive. It is, for instance, possible that divergence takes place at the level of academic discipline, while convergence occurs simultaneously at the level of institutions within a system and among systems. For the approach in this paper (which focuses on the systems level), these distinctions are relevant, because national policies might have been influenced not only by EU policies, but by higher education institutions and developments at the level of academic discipline (e.g., academic norms and values).

The question now is whether change (convergence or divergence) was caused by EU higher education policies or by other circumstances? In principle, convergence in national higher education policies might have been a consequence of:

Reactions by EU member states to policies of European institutions (the Commission and the European Court of Justice are particularly relevant in that regard);

Developments at the level of higher education institutions in the individual EU member states, as well as developments at the level of academic disciplines;

Mutual influencing of policies by individual EU member states.

Hence, reasons for convergence could have been based on:

EU-related coercion and an insight into the benefits of abiding by EU decisions[7];

An insight into the benefits of accommodating to institutional changes *within* national higher education systems, as well as an insight into the benefits of accommodating to developments inside academic disciplines;

Imitation and member state-related coercion (i.e., one state bullying another state into accepting a particular policy).

EU-related coercion and member state-related coercion, as well as insights into the benefits of abiding by particular EU decisions or of accommodating to developments at the level of higher education institutions or academic disciplines, can be subsumed under self-interest guided motivations[8]. Imitation can be based on rational self-interest, as well as on "self-prescription"[9]. Self-prescription includes non-calculative imitation of authority and non-calculative imitation of a numerically dominant example[10]. This leads us to developing the following tentative general scheme:

---

7   In the latter case we deal with benefits that are not related to coercion.
8   It will be difficult, however, to find an instance of one EU member state bullying another one into accepting a particular policy in the field of higher education.
9   I am using the term "self-prescription", instead of merely "prescription", because the latter may imply prescription by others, i.e. a form of coercion.
10   Fashion is a good example of the latter type of behaviour.

General scheme

Possible reasons for convergence of national higher education policies in the member states of the European Union include (from the perspective of national policy-makers)[11]:

Self-interest

Coercion
by EU institutions;
by other member states.
Self-interested behavior that is unrelated to coercion;
benefits related to responses to EU policies;
benefits related to developments at the level of national academic institutions;
benefits related to developments at the level of academic disciplines;
calculative imitation of other policies[12].

Self-prescription

Non-calculative imitation of the policies of an authority;
Non-calculative imitation of numerically dominant policy examples.

My next steps will be to assess whether convergence/divergence has occurred in the policies of the Netherlands, Germany and Belgium/Flanders, Great Britain, Sweden and Finland. Subsequently, I will attempt to assess if convergence/divergence (or none of the two) was the result of EU policies or of other factors. First of all, let it be noted that at the time we are concerned with, the EU had promulgated very little legislation on higher education. It had, however, passed a variety of policy initiatives in the domain of mobility (e.g., COMMETT, ERASMUS, SOCRATES and DA VINCI action programs). The Sorbonne Declaration and the Bologna Declaration were essential for the promotion of convergence processes in a number of domains, but they were not EU documents. The Sorbonne Declaration was an initiative taken by four member states (Germany, France, Great Britain and Italy). The Bologna Declaration was an initiative by a very significant number of European countries, many of which were not even EU members.

---

11   Divergence, on the other hand, can be a result of an absence of coercion at the European level, or of individual EU member states reacting differently to EU legislation. In both cases, divergence will be the consequence of EU member states adapting their policies in dissimilar ways to the environment.

12   Imitation can be based on either a rational, calculative understanding of its benefits or on motivations that cannot be reduced to rational calculation. The former type of imitation will be called "calculative imitation", the latter "non-calculative (or self-prescriptive) imitation".

Evolutionary Theory Applied to Institutions 151

Developments in the Netherlands, Belgium/Flanders, Germany, Sweden, Finland and Great Britain

Before the 1980s, the state was an essential actor in higher education policies in the Netherlands, Flanders, Germany, Sweden and Finland. Governmental authority, however, did not play the same role in the higher education systems in these countries. In the Netherlands, Belgium (until the federalization of the Belgian state in 1989), Sweden and Finland this authority was in the hands of the central government (with regard to the policy sector in question), whereas in Germany the *Länder* had the essential decision making power in higher education issues.

The Belgian higher education system was heavily dependent on the delicate Belgian balance between the different language communities, as well as the different ideological groups. Belgian politics in general, as well as educational policy in particular, were thus influenced to a significant degree by political relationships and deals (Wielemans 1991, 1). In 1959 a law was adopted which split the education system into three networks:

A public (state) network, the Minister of Education being the organizing authority;
An officially subsidized network, organized by municipalities and provinces;
A "free network", run mainly by Catholic authorities (ibid., 3).

The higher education systems in the Netherlands and Germany, on the other hand, were not characterized by this complicated ideological-political balance.

All the five mentioned systems were imbued with the ideals of equality and equivalence and all five faced a massification of higher education in the 1960s. In the 1980s all five systems gradually switched from the equality and equivalence policy to policies based on institutional competition and on the quality of the product they were delivering. In Belgium and Germany, higher vocational schools were officially introduced in 1970 (Wielemans 1991, 2-3; Van de Maat 1999, 17).

The problems that the Netherlands faced in the 1970s, and that resulted in a change of policy towards more competition in the 1980s, were a high student drop-out rate, lengthy study periods, insufficiently adequate academic staff and inefficient institutional management (Goedegebuure 1993, 190). In the late 1970s the two-tier structure was introduced, as well as retrenchment operations that resulted in the closure of some departments and the reshuffling of programs. Conditional funding for research was introduced as well (Boezerooy 1999, 11). In 1982, the funding of universities ceased to be enrolment driven and began to be based on an assessment of quality and social relevance. Instead of receiving a block grant, universities had to "earn" a part of their budget (Goedegebuure 1993, 199). In 1983, the Netherlands became the first country to adopt a formal quality assurance system. This system was called the "conditional funding policy" and it came down to institutions obtaining more autonomy if they delivered quality education (Boezerooy 1999, 50-51). Also in 1983, the Dutch Ministry of Education published the White Paper "Scale Enlargement, Task-Reallocation and Concentration" (STC). The Ministry envisioned merger processes, i.e. the emergence of multidisciplinary, medium-sized institutions with considerable autonomy in the sector of vocational education. These processes did indeed occur, even to a higher degree than the Ministry envisioned (Goedegebuure 1993, 191). The document also aimed at an increase in institutional autonomy and an increase in institutional efficiency through economies of scale (Boezerooy 1999, 12).

Developments in the German higher education system in the mid 1980s followed many Dutch trends of the first half of the 1980s. In 1983, the Federal Ministry of Education came up with the idea of more differentiation and competition (Frackmann 1993, 151). The new 1985 Framework Act (*Rahmengesetz*) provided higher education institutions with even more autonomy than they had on the basis of the 1976 Framework Act (according to which institutions provided input into state level planning) (ibid., 143). The planning sections from the 1976 law were entirely removed in 1985 (ibid., 143). According to the 1983 Act, institutions were allowed to propose new programs, but the ultimate authority was with the state Ministry of Education. Final decisions concerning curricula were within the institutions (ibid., 143). In 1985, the German Science Council (*Wissenschaftsrat*) followed the recommendation of the Ministry of Education and decided that differences in quality and performance should be made visible and that funds should be allocated competitively (ibid., 151). This decision corresponded to the in 1983 adopted Dutch quality assurance system and to the Dutch idea of "conditional funding". Hence, changes in the German higher education system of the mid 1980s were in line with similar developments in Holland from a few years before. These changes took place in the domain of the idea to establish more competitive, market-oriented, efficient higher education systems, in an increase of institutional autonomy, as well as in the domains of quality assurance and funding.

In Belgium, the relationship between the government and the higher education system underwent some changes as well in the mid 1980s. In 1986, the Saint-Ann austerity plan was adopted, as well as a plan according to which the non-university sector was to be rationalized. Both plans were promulgated by the Ministry of Education (Wielemans 1991, 4).

The second half of the 1980s was not marked by major developments in the field of higher education in any of the analyzed countries, except for one essential event in Belgium. Its relevance, however, exceeded the higher education sector. In 1989, the Belgian state was federalized and higher education policies were almost com-

Evolutionary Theory Applied to Institutions 153

pletely transferred to the level of regions and language communities. Policies regarding research were somewhat less affected (ibid., 2). Local boards of administration were set up in all non-university institutions of higher education (ibid., 4). From this point we will focus on the Flemish system of higher education.

In the 1990s, the higher education systems of the Netherlands and Flanders continued to converge with regard to a number of essential issues, whereas the German higher education system did not experience major changes (Frackmann 1993, 160). In 1990, HBO (*Hoger Beroepsonderwijs* – Higher Vocational Education) institutions in the Netherlands started with contract research activities (Boezerooy 1999, 19). The "HOOP document" of 1992 acknowledged the importance of the report on skills shortages in Europe by the advisory Committee of the European Commission (IRDAC, 1991) and supported its most important recommendations, which included an increase in graduates in science and technology, more investment in recurrent education and an improvement of the productivity of education systems. This policy position can also be found in the ministerial budget statements of 1992 and 1993. The Higher Education and Research Act of 1993 (based on the 1985 HOAK document) also insisted on institutional autonomy and a governmental non-intervention policy (ibid., 13). Institutions were to obtain substantial freedom of programming. The quality of a new program was to be judged *ex post* (Goedegebuure 1993, 1998). In general, the quality assessment system expanded considerably in the 1990s and was based both on self-evaluation as well as external evaluation (by the Inspectorate of Higher Education or by peer review) (ibid., 208). In terms of funding, the Dutch higher education institutions acquired their financial means from three sources: directly from the Ministry (some 73%), from the Dutch Research Council (some 5%) and from contract research. The trend was that contract research funding was on the increase (Boezerooy 1999, 31). Tuition was also required and was equal for Universities and the HBO sector (ibid., 31). Since 1999 the specific Dutch system of *numerus fixus* is also being transformed: all students who have 8.00 or more as their G.P.A. would be automatically admitted to University programs they wish to enrol in (ibid., 20). Finally, Holland adopted the (Anglo-Saxon) mode of the BA/MA degree structure (Kamerman 1999).

In Flanders a governmental decree was issued in 1991, according to which vocational education (HOBU) was to obtain much larger autonomy, whereas less central regulation and more "steering from a distance" was prescribed for Universities (Beverwijk 1999, 33). The 1991 decree also dealt very seriously with quality assurance. The Flanders Inter-University Council (VLIR) was established and supposed to coordinate quality assurance, based on internal and external control (external control through peer review and without government intervention). The VLIR was to cooperate closely with a similar body in the Netherlands

(VSNU) and used its reports as a guideline (ibid., 45). The decree reformed the academic personnel structure also substantially, making it not much different from the Dutch, British or US structure (assistant professor – associate professor – full professor). A more flexible system of transitions between different training levels was introduced (ibid., 11). In general, changes in the Flemish system of higher education in the 1990s included a reduction of government involvement (including a reduction in government subsidies and more competition (Wielemans 1991, 10)). Apart from these general trends, convergence with the Dutch system took place on the level of quality assurance, on the level of entrance requirements, and possibly on the level of higher education structure (the Netherlands adopting the in Flanders already existing BA/MA degree structure).

The German higher education system, though converging with the Dutch and Belgian/Flemish in the 1980s, did not change substantially in the 1990s and remains different from the systems in the two Benelux countries in, for instance, its specific federal structure with decentralized authority, its insistence of the primacy of equality and equivalence (something which in Belgium was assumed in light of its specific ideological-political balance, but that was being gradually suppressed), the absence of tuition fees (Van de Maat 1999, 42), open access policies in Universities (at least in principle) (Frackmann 1993, 135), as well as the absence of a national quality assessment system (Van de Maat 1999, 47). On the other hand, the Conference of Rectors and Presidents of Universities adopted in 1995 a resolution with special reference to assessment of teaching (both internal and external assessment), whereas in 1997 consultations took place both at the level of *Länder* and at the inter-regional level about the adoption of a national quality assessment system (ibid., 47). Another trend in the direction of convergence was the increase in importance of higher education funding from the private sector (ibid., 40).

Part of the distinctiveness of the Finnish higher education system is largely a product of early twentieth-century Finnish nationalism. The salient features of this system were the high proportion of women, people from less privileged social backgrounds, as well a high proportion of the general population attending higher education, a delayed expansion of technology and business science, and, last but not least, a significant growth in higher education (especially in the natural sciences and humanities) prior to the full impact of industrialization (Kivinen/ Rinne 1996, 102). Since the 1950s enrolment in higher education increased sevenfold, which was followed by the introduction of entirely new fields and subjects (ibid., 101).

During the 1970s, the higher education system in Finland was being strongly centralized. It became structurally homogenous: the state became the owner of all higher education institutions, major reforms were carried out concerning curricula, degrees and the internal administration of the universities, the structure of degrees was standardized, while a non-university higher education sector

Evolutionary Theory Applied to Institutions 155

remained absent (ibid.: 100-101, 110). This homogeneity remained unchanged during the 1980s. In the 1990s, however, many previous reforms were undone: the state ceased to be the sole owner of the institutions, the ideal of standardization was replaced by diversification as the dominant trend, whereas a non-university higher education sector began to emerge (ibid., 110). Traditional demands for equality were replaced by demands for greater efficiency (ibid., 103).

In 1994, the Finnish Ministry of Education concluded that the status of vocational education in Finland (which is somewhere between secondary and higher education) does not correspond to developments in other countries, and stated: "The structure of the Finnish system is obscure. The polytechnics are to be built up on the foundation provided by the current vocational education system. At a time of growing international cooperation and European integration, advanced vocational education should evolve into a system with an identity of its own, forming a distinct non-university sector of higher education" (Ministry of Education, 1994). Access to polytechnics ("AMK institutions") is open to those who have taken the Matriculation examination, have completed upper secondary school, or have a vocational basic qualification (or post-secondary qualification) or a corresponding international or foreign qualification (Beverwijk/Schrier 1999, 17). Student selection was based on school achievement, work experience and (in a significant number of cases) entrance examinations (ibid., 17). Access to Universities was open (ibid., 23). Another difference between Universities and AMK institutions was in the manner in which they were governed: Universities were run by the national government, AMK institutions by local authorities or privately (ibid., 37). The principle of equality still appeared to be preserved in the fact that there were no tuition fees[13] (ibid., 38). Funding of higher education institutions was based on lump sums administered by the government (ibid., 35).

In the domain of quality assurance, the Finnish authorities became active in 1986. In that year, the government decided to require from all universities to introduce assessment systems with compatible information about the results of research and instruction (ibid., 49). In 1996, The Finnish Higher Education Evaluation Council (FINHEEC) was established. It was supposed to advise the Ministry of Education and to assist higher education institutions with self-evaluation (ibid., 49). In general, evaluations of higher education institutions consisted of three stages: self-evaluation, external evaluation and a final report from this (ibid., 50).

Finally, it deserves mention that the degree structure in Finland had changed as well. In the 1970s, the first degree was established as a standard at the Master's level. Since the 1990s Finland has the BA/MA structure (ibid., 21).

---

13 The Finnish government's proposal to introduce tuition fees was not accepted by parliament, which alleged that the proposal violated the principle of equal opportunities (ibid., 35).

In Sweden, the higher education system was reformed in 1977 on the basis of a 1973 parliamentary report. Unlike the Finnish higher education system that was being centralized during the 1970s, the Swedish system was reformed in the direction of a decentralization of authority from the national to the regional and local level, as well as in the direction of more flexible funding schemes (Klemperer 1999, 15). In addition to these changes, the 1977 reform resulted in the development of new institutional classifications[14], a reorganization of the institutional structure and of access to higher education (limited access, determined by government, and more places for students above 25), as well as a new organization of programs, courses and the credit system (Svanfeldt 1994, 241). The 1977 reform also aimed at helping people with working class backgrounds to get access to the higher education system (Klemperer 1999, 16).

In 1983, the Swedish higher education system was being reformed again. This reform consisted of the following: a further decentralization, an increase in the autonomy of higher education institutions (although representatives of external interests were supposed to be a substantive part of the governing boards of the institutions), as well as of a move towards less bureaucracy (Svanfeldt 1994, 241).

In 1993, a law was being prepared that would give vice-presidents and local boards much greater freedom to administer their University or college, reform the degree system (reducing the existing 120 programs to some 47 degrees) and allow some institutions to become private NPOs (ibid., 251-252). This reform was in line with the idea of developing competition among the institutions (ibid., 251). In practice, the 1993 law resulted in a number of changes in the Swedish higher education system. The curricula of individual programs were to be determined by individual institutions instead by the Ministry of Education. Institutions were given the freedom to distribute the financial resources granted to them by the government. The allocation of grants among institutions became to be determined by student demand and by the achievements of the individual institutions. The organization of study and the range of courses was to be decided locally rather than nationally. Students were given the freedom of choice concerning their study paths on the basis of a new internationally valid Degree Ordinance, attached to the 1993 Higher Education Ordinance. Finally, the Office of the University Chancellor (*Kanslersambetete*) was established with the aim to assess the quality of institutions and programs, as well as to examine the right of institutions to award degrees (Klemperer 1999, 16)[15].

---

14 Prior to 1977, higher education was divided into four sectors: universities, colleges, institutes and vocational schools. After the reform, Sweden adopted an integrated unitary system of higher education (Svanfeldt 1994, 241).

15 In 1995, this Office was abolished and its role taken over by a new organization which is supposed to read self-evaluation reports and to employ special audit teams (Klemperer, 50). In the

Similar to the funding of higher education in Finland, the Swedish Parliament began to allocate lump sums to institutions, which were supposed to divide them as they considered appropriate (Svanfeldt 1994, 255-259). Government funding of Universities, however, became much more extensive than that of vocational schools (*hogskolan*), which thus had to sell courses to companies and other organizations (ibid., 258). As in Finland, students in Sweden were not supposed to pay tuition fees, since the still existing principle of equality was considered to imply that all students who need assistance from the central government were supposed to obtain it (Klemperer 1999, 39).

General admission to Swedish higher education institutions was determined by the parliament and government (Svanfeldt 1994, 245-246). After the reforms, individual institutions obtained the right to decide themselves about *specific* requirements for admission to programs (e.g., about how many and which students to admit). This could lead to a situation in which the central admissions system would become a merely coordinating one (ibid., 248). General criteria for student selection were: 1) that the student has at least completed upper secondary school, adult secondary school, folk high school, foreign secondary school, or that (s)he has reached the age of 25 with four years of previous work experience, and 2) that the student is sufficiently proficient Swedish and English (Klemperer 1999, 20).

The uniformity of the higher education system in Sweden decreased with the passing of time, while the principle of competition became increasingly prominent. The integrated system of higher education that was developed after 1977, had changed substantially. *Hogskolan*, for instance, developed a profile of their own.

The degree structure in Sweden differed from the BA/MA structure. Although both degrees do exist in Sweden, they were preceded by another one (the *diploma*). This made the Swedish situation in this domain relatively unique (ibid., 18).

When comparing the Finnish and Swedish system of higher education, one can conclude that they became similar even before Bologna, but that they converged via different paths. In both systems, the state traditionally played in important role. The Swedish system, however, experienced decentralization in the 1970s, while the Finnish system became less centralized only in the 1990s. In Finland of the 1970s and 1980s, the state was the owner of all higher education institutions, degrees were standardized, while the non-university sector remained absent. The 1977 reform in Sweden, on the other hand, resulted in decentralization, while the 1983 reform increased the autonomy of higher education institutions. With the passing of time, higher vocational education also developed a

---

1980s, quality assessment took generally place on an *ad hoc* basis, while studies on the evaluation of institutions began to emerge in the late 1980s (ibid., 49).

profile of its own in Sweden. In the meantime, no significant changes occurred in the centralized and government controlled system of higher education in Finland.

In the 1990s, however, important changes in the direction of more competition and decentralization did take place in Finnish higher education. The state ceased to be the sole owner of higher education institutions, standardization was replaced by diversification, and a locally or privately run non-University higher education sector emerged (the AMK institutions). The principle of equality remained present in the fact that access to Universities (not to AMKs) remained open and that tuition fees were not introduced. The BA/MA structure was introduced in Finland, while Universities obtained financial resources from the government in the form of lump sums. The system of quality assurance developed in Finland after 1986 with the establishment of the FINHEEC.

In Sweden, the reforms of 1993 introduced more competition, some privatization of higher education institutions, as well as a system of quality control. These changes made the Finnish and Swedish higher education systems rather similar. In both countries, diversification was on the rise, the state ceased to be the sole owner of higher education institutions, the non-University sector gained in importance, higher education institutions were financed in a rather similar manner (lump sums from the national government to the University sector, less funds for *hogskolan* and AMKs respectively, no tuition fees), while entrance to higher education was also based on similar criteria. The system of quality control developed in both countries as well, and was operationalized in a similar manner by the FINHEEC and *Kanslersambetete* (until 1995) respectively. The degree structure in Finland and Sweden was not identical, but relatively similar as well.

It would be interesting to understand why convergence between the Finnish and Swedish systems of higher education occurred via different paths. The Finnish system proved to be very open to European influences (Ollikainen 1999, 5). The attitude of the relevant institutions in Finland (the government, Ministry of Education, expert bodies, interest organizations and educational institutions) was highly favorable towards E.U. higher education policies (ibid., 6). This positive attitude was most pronounced at the highest political level (ibid., 7). Its impacts had been most visible in the domain of international cooperation. Indirect consequences of European integration, however, influenced also the structure and content of Finnish higher education (ibid., 13).

The impact of the European Union on Swedish higher education appears less pronounced. As a matter of fact, Sweden began reforming its centralized system before Finland and before joining the European Union. It seems also that changes in Swedish higher education were more influenced by changes in government at the national level. It is interesting to note in that regard that the turn of Swedish higher education toward more competition occurred with the installation of a

Evolutionary Theory Applied to Institutions 159

center-right government in the early 1990s, while another turn *away* from the market occurred with the Socialists coming to power again in the mid-1990s.

Higher education institutions in the United Kingdom were traditionally marked by a high degree of institutional autonomy[16]. The role of government in the higher education system was limited (Brennan/Shah 1994, 290). When the demand for higher education began to rise in the 1950s, the higher education institutions responded with raising entry requirements. This led to tensions inside and outside the system. In the sixties, a Committee (the Robbins Committee) was established with the objective to investigate the future of higher education in Great Britain. It published a report in 1963 that insisted on the principle of equality: "all young persons qualified by ability and attainment to pursue a full time course in higher education should have the opportunity to do so" (ibid., 292). Contrary to the recommendations of the Robbins Committee, however, the expansion of higher education did not take place in the University sector, but through the development of polytechnics and colleges (ibid., 292-293).

In 1985, a report was published (the Jarratt Report) which contained proposals for a transformation of university management practices. The proposals were very much in line with the concept of university as corporation. The report indicated, among else, that vice-chancellors should be chief executives and that more corporate planning should replace the power of the departments (ibid., 299). For the polytechnics, a similar document was presented in the same year by the National Advisory Body's Good Management Practice Group. This document, however, insisted on a lower degree of centralization of institutional control, and emphasized the role of departments, sub-units and individuals in it (ibid., 299).

With the Education Reform Act of 1988, the polytechnics and most of the colleges were removed from the control of local education authorities, and tenure was removed from Universities. With the same Act, the Polytechnics and Colleges Funding Council were created, as well as the Universities Funding Council (replacing the old University Grants Committee). Both new funding councils were responsible to the government's Department of Education and Science, which resulted in convergence between the university and polytechnics/colleges sector (ibid., 297). Furthermore, funds were not given anymore as a block grant. Instead, they were separated and handed over to research councils to which applications were supposed to be made on an *ad hoc* basis. The funding councils introduced a comparative element in making their decisions, based on considerations relating to student demand, price and quality (ibid., 300).

---

16 There are substantial differences between the English, Scottish and Welsh higher education systems (Brennan and Shah 1994: 290). I will concentrate on the English system, because of its comparative relevance.

The government published in 1991 a White Paper, "Higher Education: A New Framework", which became legislation the following year. It allowed the polytechnics to award their own degrees and the right to adopt the title of "University" (ibid., 296)[17]. In 1992, polytechnics were granted university titles, which meant the end of the binary divide (ibid., 293).

Universities in the U.K. were funded by the Higher Education Funding Council (this applied primarily to teaching, but also to research, and it was the responsibility of the Ministry of Education), by Research Councils (this applied to research only, and it was the responsibility of the Department of Trade and Industry), and by private sources (Beverwijk 1999: 29). An increasing number of funds came from non-core funding. This led a number of observers to talk about the emergence of an "intellectual proletariat" in Britain (Brennan and Shah 1994: 309). Tuition fees used to be high up to 1997/1998, but were paid by the Local Educational Authorities (in the case of British full-time students). Since the 1998-1999 academic year, however, full-time students were generally charged 1,000 pounds annually. Students from low-income families were partially exempted from paying tuition fees (Beverwijk 1999, 47-48).

The legislation resulting from the White Paper of 1991 created also a new quality assurance agency (the Higher Education Quality Council – HEQC). Funding councils could also undertake quality assessment at the program level (Brennan/Shah 1994, 297). The Further and Higher Education Act of 1992 stated that quality control is not only possible through the market, but also through institutions and the newly formed HEQC (ibid., 305). Also, funding councils should take into account quality in making their decisions (ibid., 305)[18]. In 1997, the Quality Assessment Agency was established. Its aim was to work together with higher education institutions on quality assurance and the development of a common terminology and credit system in higher education (Beverwijk 1999, 59).

The system of admissions in the United Kingdom became very selective. Only 40% of the candidates were admitted. Because of their autonomous status, Universities had a lot of freedom in deciding who to admit. (ibid., 21). In general, admission criteria consisted of exam results, references, personal arguments and motivation (ibid., 22).

Finally, it is in order to mention that almost all UK Universities were public (the only exception is the University of Buckingham), but that there was no national curriculum. The latter phenomenon was the consequence of the fact that

---

17 Prior to the White Paper, degrees were awarded by the Council for National Academic Awards (Brennan/Shah 1994, 298).
18 The reports of funding councils are based on institutional self-assessment, statistical indicators and external reviews (Brennan and Shah 1994, 312).

the Senate of Universities had the exclusive power to approve new courses and programs (Brennan/Shah 1994, 294-297).

When comparing the situation in British higher education with developments in the other five selected countries, one will notice convergence in terms of an overall trend toward the domination of market mechanisms, but also a number of essential differences. The non-University sector in the United Kingdom began to develop already in the 1960s, but in 1991 polytechnics acquired the status of Universities. In that regard, therefore, one will notice divergence. Other important differences between the British system and the other five, included funding mechanisms (e.g., students did have to pay tuition in Britain, while non-core funding also played a more important role in Britain than in the other countries), as well as the admissions system (the British being the more selective one). Convergence between the six systems, on the other side, had also taken place in the domain of the emergence of quality assurance systems, decentralization (in Britain the polytechnics system was being decentralized in 1985), and the development of similar degree structures (to a lesser extent in Sweden).

In conclusion: the Dutch, Flemish, German, Finnish and Swedish systems of higher education converged in terms of structure, funding and quality assurance, whereas the system in Great Britain converged with the other five with regard to some structural aspects (the degree structure to some extent – the abolishment of polytechnics in Britain, however, being an example of divergence), as well as the development of systems of quality assurance. The funding of higher education institutions in Great Britain remained different. In terms of student and teacher mobility, EU policies had a direct impact on all six states. Finally, it deserves mention that up to the Bologna Declaration the effects of European policies on British higher education were limited. One of the reasons may be the fact that Britain was until then ruled by Conservative governments, which were not too favorably disposed to European integration.

In all six analyzed higher education systems, one can observe a trend toward the principles of competition in the period we are concerned with. European integration, together with the process of globalization, played an essential role in the development of this trend. Higher education institutions became less dependent on the state, functioned more autonomously, and were increasingly subjected to "natural selection". This was an overall trend in all six states (although there were periodical increases in government involvement), and it can be concluded that convergence of national higher education systems and policies has outweighed divergence in the six countries in the period under investigation. This convergence resulted in its institutional formulation: the Sorbonne Declaration (1998) and the Bologna Declaration (1999).

The highly important impact of national higher education policy making on the increasingly competitive European environment has also been assessed. Governments in the six analyzed countries had indeed reacted to European integration and globalization with creating a more favorable environment for increasingly competitive higher education systems. The government of the United Kingdom has done that even earlier. This trend indicates that it is warranted to speak about a "snowball effect" in the six countries. The emergence of an increasingly competitive environment leads policy makers to create favorable conditions for institutions to compete in this environment, which results in an even stronger role of competition. In light of this effect, it is clear why convergence outweighed divergence in higher education policies of EU member states. That has indeed occurred in all six analyzed states.

European integration resulted in a decrease of environmental diversity and a more salient role of the principles of competition. National policies reacted to this by creating more favorable conditions for competition, which led to a further decrease of diversity. Furthermore, policy examples were imitated, while the role of isomorphism was augmented by the high degree of professionalization of the sector in question[19].

It thus appears that convergence in the higher education policies and sectors did occur in the analyzed countries in the 1980s and 1990s, but the role of the European Union in that regard was only minor. Specifically, the European Union hardly adopted legislation that forced any of the six countries to make changes in their higher education systems. Thus, there was barely any direct impact of EU higher education policies in the six member states in question. The only exception in that regard was the domain of mobility, which was stimulated by the EU through a variety of action programs (COMMETT, ERASMUS, SOCRATES, LEONARDO DA VINCI etc.). In terms of indirect impact, however, it seems that the mentioned EU action programs in the domain of mobility impacted on changes in the structure of higher education systems (e.g., an insight into the convenience of having similar BA/MA degree structures may translate into relevant institutional changes), as well as on changes in the domain of quality assurance (e.g., similar quality assurance criteria among the member states will positively influence student mobility).

Convergence in higher education policies in the six countries may certainly be explained by authorities imitating each other's policies. Behind this imitation, however, appeared to be self-interest. The relevant governments judged what the dominant trends would be in Europe in the future and generally avoided to lag

---

19 For my analysis of competition and cooperation processes between private and state universities in one specific non-EU country (Serbia), see Rakic (2006).

Evolutionary Theory Applied to Institutions 163

behind with their policies. In that regard, the European Union was an integrating factor and the authorities of the member states were generally aware that, even if they were not directly coerced into a particular policy by the EU, major deviances from the dominant trend would have a negative impact on their country. It should also be emphasized that the ensuing important documents in the field of higher education (the Sorbonne Declaration and the Bologna Declaration) were initialized bottom-up from the member states. In fact, Sorbonne and Bologna appear to have been both a *reaction* to convergence processes that have taken place among higher education policies of European states, as well as an *instrument* for promoting further convergence. All in all, the EU did not have a central role. Circuitously, EU policies had an impact. Thus, the dominant motives behind convergence processes have to be sought in benefits related to EU policies, as well as the member states' calculative and possibly non-calculative imitation of each other's policies[20].

Link to Theories of Institutional Isomorphism

The situation in which government institutions imitate each other's policies introduces us to literature on institutional isomorphism. In this chapter I will firstly position the theoretical model from the "general scheme" in relation to two pivotal works dealing with institutional isomorphism: DiMaggio and Powell's article "The Iron Cage Revisited: Institutional Isomorphism and Collective Rationality in Organizational Fields" (1983) and Richard Scott's book *Institutions and Organizations* (1995). Afterwards I will demonstrate why our model can fully account for the problem that was raised in this paper, and consequently does not have to be upgraded with either DiMaggio and Powell's article or Scott's book.

"The Iron Cage Revisited: Institutional Isomorphism and Collective Rationality in Organizational Fields"

According to the authors, rational actors make their institutions increasingly similar, while attempting to change them. This convergence is based on three isomorphic processes: coercive, mimetic and normative (DiMaggio/Powell 1983, 147). Coercive isomorphism results from formal and informal pressures on one organization by another organization which is dependent upon it and by "cultural expectations in the society within which organizations function" (ibid., 150). Mimetic isomorphism is encouraged by uncertainty, "when organizational technolo-

---

20 These three motives are classified under A2a, A2d, B1 and B2 in the "general scheme".

gies are poorly understood, when goals are ambiguous, or when the environment creates symbolic uncertainty" (ibid., 151). Normative isomorphic change is a consequence of professionalization, i.e. of "the collective struggle of members of an occupation to define the conditions and methods of their work" (ibid., 152)[21].

Two major differences appear between the previously presented "general scheme" and DiMaggio and Powell's theory. First, the "general scheme" deals with *reasons* for convergence, whereas DiMaggio and Powell have *mechanisms* of convergence in mind. Second, DiMaggio and Powell address exclusively rational mechanisms, while I believe that non-rational (or rather non-calculative, to avoid confusion with "irrational") mechanisms may also lead to certain policy outcomes, including the direction of institutional change[22]. Thus, "calculative imitation" in the general scheme can be a motivational background for all three types of isomorphic processes in DiMaggio and Powell.

"Institutions and Organizations"

Scott also distinguishes DiMaggio and Powell's three mechanisms of isomorphic change, but regards them as elements of three *pillars of institutions* as well[23] (Scott 1995, 35). He differentiates the regulative, normative and cognitive pillar. The regulative pillar includes rule-setting, monitoring and sanctioning activities (ibid., 35). The normative pillar relates to rules that introduce a prescriptive, evaluative and obligatory dimension into social life. In Scott's words: "Actors conform not because it serves their individual interests, narrowly defined, but because it is expected of them; they are obliged to do so" (ibid., 39). The cognitive pillar emphasizes stabilizing effects of shared definitions of social reality (ibid., 40).

It seems, however, that the normative and cognitive pillar do not differ much from each other. They both pertain to a particular moral obligation, with the difference that "cognitivists" do not tend to attach an absolute (or even objective) value to their morality. In that sense, the cognitive pillar can be regarded as a relativized normative pillar. In contrast to the normative and cognitive pillar, the regulative pillar does not pertain to moral obligation, but to a more coercive element in obligation. Thus, the regulative pillar stimulates compliance on the

---

21 The authors note that universities and professional training institutions are important for the development of organizational norms (DiMaggio/Powell 1983, 152). Normative isomorphism may, therefore, be an important explanatory mechanism in research on the impact of academic institutions and disciplines on national (and supranational) policy making in the field of higher education.
22 Fashion is an example of isomorphism that is mostly based non-calculative imitation.
23 The three pillars are identified as "making up or supporting institutions" (Scott 1995, 35).

basis of self-interest, the normative and cognitive pillar on the basis of what I called self-prescription.

We have seen that the *direct* role of the European Union in convergence processes in the six analyzed countries was only present in the domain of mobility. Its *indirect* role could be noted in other domains as well. For instance, an increase in student mobility within the European Union had an impact on the development of similar programs or similar degree structures in the member states, as well as on the development of comparable quality assurance mechanisms. In all this, calculative imitation of each other's policy examples by member states was the dominant mechanism of institutional isomorphism. Apart from this calculative imitation, it is also possible that non-calculative motivations were behind imitation, i.e. that policy makers in a particular state imitated the policies of other states on the basis of moral authority or numerical domination, rather than on the basis of pure calculation. Thus, it was a (soft) form of coercion by the European Union that lead to convergence in the domain of mobility (with side-effects in other domains), and calculative (and possibly also non-calculative) imitation of policy examples that resulted in convergence in quality assurance and the structure of higher education systems in the six member states. All things considered, it turns out that the issue of interest here can be fully explained by the theoretical model from the "general scheme", and that it is thus unnecessary to seek additional explanatory power in the works of DiMaggio/Powell and Scott.

It remains to be clarified why convergence in the higher education systems in the six countries was on the increase in the period we dealt with, while similar developments were not present to that degree before. It appears that the policies of the European Union were a significant contributing factor in the analyzed time-frame. In addition to that, it is important to note that a massification of higher education in the 1960s and 1970s resulted in governments being less able to spend sufficient financial resources on higher education. This development was augmented by the economic crisis of the 1970s and 1980s. The consequence was an increasing orientation of higher education institutions toward the market[24]. Since this orientation became increasingly dominant throughout the European Union, it is understandable that events resulting from it led to an overall trend toward convergence in a number of domains of the higher education systems in the member states. Hence, it appears that it were both the consequences of a massification of higher education in Western Europe, as well the increasingly important role of the European Union, which contributed to significant develop-

---

24 This process became even more pronounced in those countries in the European Union in which the center-right took office.

ments in the direction of convergence in the period in question. It has been demonstrated, however, that the essential mechanism responsible for convergence was institutional isomorphism. It is the survival mechanism of institutional imitation (both calculative and non-calculative), rather than direct EU policies, that accounts for observed convergence of higher education policies of European states in the period leading to Sorbonne and Bologna.

Epilogue: Repercussions of Institutional Adaptation for Evolutionary Theory

Let me close this paper with the announced brief excursion that is relevant for evolutionary theory in general. It has been demonstrated in the previous sections that the concept of adaptation mechanisms that is derived from evolutionary theory can explain convergence of higher education policies. Hence, institutional theory can borrow concepts from evolutionary theory. Is a reverse process also possible? In other words, can evolutionary theory borrow from institutional theory?

One important argument in the various debates on evolutionary theory remains that even if life can be explained without the idea of "intelligent design" (and humanity might find such an explanation if it is to survive for a sufficiently long time) it is still not a proof that such design does not exist. It is imaginable that God could have created the world with its laws of development and have let it continue in time[25]. Similarly, the discovery of adaptation mechanisms of higher education policies to an increasingly competitive and Europeanized environment (substantiated by their convergence[26]) does of course not imply that these policies do not have a "creator".

It has been shown that some higher education policies have emerged as a consequence of institutional imitation (calculative or not). This finding, however, does not have to lead to the conclusion that these policies exist independently from concrete political actors. Hence, relevance ought to be attributed both to self-regulating mechanisms of the market, to Europeanization, as well as to policies that have been developed by specific political actors. This conclusion can shed some additional light on the alleged incompatibility of evolutionary theory with the idea of intelligent design, i.e. it can serve as an argument *in favor of*

---

25 Instructive for this line of argumentation is Ivanovic (2009, 380).
26 As already noted, convergence does not have to be the only indicator of adaptation at work. Evidence of adaptation mechanisms can also be divergence of higher education policies – in the case that some policies attempt to find a "niche" for themselves in the market. It is dubious, however, whether cogent evidence of this type has been found.

Evolutionary Theory Applied to Institutions 167

their compatibility. The fact that this light comes from a seemingly unlikely source, i.e. from institutional theory, ought not to diminish its usefulness. Moreover, it shows that the answer to our last question is affirmative: evolutionary theory *can* successfully borrow concepts from institutional theory.

Bibliography

Beverwijk, J. (1999): Higher Education in the United Kingdom: Country Report of the CHEPS Higher Education Monitor. CHEPS, Enschede.
Beverwijk, J./Schrier, E. (1999): Higher Education in Finland: Country Report of the CHEPS Higher Education Monitor. CHEPS, Enschede.
Beverwijk, J./Lange, S. de (1999): Higher Education in Flanders: Country Report of the CHEPS Higher Education Monitor. CHEPS, Enschede.
Boezerooy, P. (1999): Higher Education in the Netherlands: Country Report of the CHEPS Higher Education Monitor. CHEPS, Enschede.
Birnbaum, R. (1983): Maintaining Diversity in Higher Education. Jossey-Bass, San Francisco.
Brennan, J./Shah, T. (1994): Higher Education Policy in the United Kingdom. In: Goedegebuure, L. et al. (eds.): Higher Education Policy: An International Comparative Perspective. Pergamon Press, Oxford.
Clark, B.R. (1983): The Higher Education System. University of California Press, Berkeley (Ca) and London.
Clark, B.R. (1996): Diversification of Higher Education: Viability and Change. In: Meek, L. (ed.): The Mockers and the Mocked: Comparative Perspectives on Differentiation, Convergence and Diversity in Higher Education. Pergamon Press, Oxford.
DiMaggio, P.J./Powell, W.W. (1983): The Iron Cage Revisited: Institutional Isomorphism and Collective Rationality in Organizational Fields. In: American Sociological Review 48, 147-160.
Frackmann, E./Weert, E. de (1993): Higher Education Policy in Germany. In: Goedegebuure, L. et al. (eds.): Higher Education Policy: An International Comparative Perspective. Pergamon Press, Oxford.
Friedman, T. (1962): Capitalism and Freedom. The University of Chicago Press, Chicago and London.
Goedegebuure, L. et al. (eds.) (1994): Higher Education Policy: An International Comparative Perspective. Pergamon Press, Oxford.
Goedegebuure, L./Van Vught, F. (eds.) (1994): Comparative Policy Studies in Higher Education. Lemma, Utrecht (Holland).

Huisman, J. (1995): Differentiation, Diversity and Dependency in Higher Education. Lemma, Utrecht.
Huisman, J. (1997): De Regulering van het Opleidingsaanbod: Een Slingerbeweging tussen Overheidsplanning en Zelfregulering. In: Beleidswetenschap 11(2), 122-142.
Ivanović, M. (2009): O Božijoj egzistenciji – staro i novo. In: Jerotić, V./ Ivanović, M. (eds.): Religija između istine i društvene uloge. Dereta, Beograd.
Kamerman, S. (1999): Bachelor/Master gaat het maken. In: NRC Handelsblad of 10 September 1999, p. 3.
Kivinen, O./Rinne, R. (1996): The Problem of Diversification in Higher Education. Countertendencies Between Divergence and Convergence in the Finnish Higher Education System Since the 1950s. In: Meek, V.L. et al. (eds.): The Mockers and the Mocked: Comparative Perspectives on Differentiation, Convergence and Diversity in Higher Education. Pergamon Press, Oxford.
Klemperer, A. (1999): Higher Education in Sweden: Country Report of the CHEPS Higher Education Monitor. CHEPS, Enschede.
Maassen, P./Huisman, J. (1999): Higher Education and European Integration (manuscript).
March, J.G./Olsen, J.P. (1989): Rediscovering Institutions: The Organizational Basis of Politics. The Free Press and Collier Macmillan Publishers, New York and London.
March, J.G./Olsen, J.P. (1998): The Institutional Dynamics of international Political Orders. Arena, Oslo.
Ministry of Education of Finland (1994): Higher Education Policy in Finland. Helsinki.
Neave, G./Van Vught, F. (eds.) (1991): Prometheus Bound: The Changing Relationship Between Government and Higher Education in Western Europe. Pergamon Press, Oxford.
Neave, G. (1996): Homogenization, Integration and Convergence: The Cheshire Cats of Higher Education Analysis. In: Meek, L. (ed.): The Mockers and the Mocked: Comparative Perspectives on Differentiation, Convergence and Diversity in Higher Education. Pergamon Press, Oxford.
Neave, G. (1996): Homogenization, Integration and Convergence: The Cheshire Cats of Higher Education Analysis. In: Meek, L. (ed.): The Mockers and the Mocked: Comparative Perspectives on Differentiation, Convergence and Diversity in Higher Education. Pergamon Press, Oxford.
North, D.C./Thomas, R.P. (1973): The Rise of the Western World: A New Economic History. Cambridge University Press, London.
Ollikainen, A. (1999a): The Single Market for Education and National Policy. Research Unit for the Sociology of Education, Turku (Finland).

Ollikainen, A. (1999b): The Single Market for Education and National Education Policy. Paper presented at the CHEPS Seminar at the University of Twente in Enschede on 4 February.
Rakic, V. (2006): Privatni univerziteti u Srbiji i Bolonjska deklaracija – takmičiti se ili sarađivati. In: Bolonjski proces i visoko obrazovanje u Srbiji. Cicero, Beograd.
Rakic, V. (2002): Evropska integracija i visoko obrazovanje: kretanja prema konvergenciji i njen uticaj na Srbiju. In: Javna uprava 1 (1), 63-80.
Rakic, V. (2001): Converge or not Converge: The European Union and Higher Education Policies in the Netherlands, Belgium/Flanders and Germany. In: Higher Education Policy 14 (3), 225-240.
Scott, R.W. (1995): Institutions and Organizations. Sage Publications, Thousands Oaks (Ca), London, New Delhi.
Svanfeldt, G. (1994): Higher Education Policy in Sweden. In: Goedegebuure, L. et al. (eds.): Higher Education Policy: An International Comparative Perspective. Pergamon Press, Oxford.
Teichler, U. (1997): The Relationships between Higher Education Research and Higher Education Policy and Practice: The Researchers' Perspective. Keynote Speech at the UNESCO Round-Table at the University of Tokyo, Tokyo.
Van de Maat, L. (1999): Higher Education in Germany: Country Report of the CHEPS Higher Education Monitor. Enschede: CHEPS.
Van der Wende, M. (1999): The Bologna Declaration: Enhancing the Transparency and Competitiveness of European Higher Education. Paper presented at the Fourth Annual Conference of GATE in Melbourne (Australia).
Van Heffen, O. et al. (eds.) (1999): Overheid, Hoger Onderwijs en Economie: Ontwikkelingen in Nederland en Vlaanderen. Lemma, Utrecht (Holland).
Van Vught, F. (ed.) (1989): Government Strategies and Innovation in Higher Education. Jessica Kingsley Publishers, London.
Van Vught, F. (1996): Isomorphism in Higher education? Towards a Theory of Differentiation and Diversity in Higher Education Systems. In: Meek, L. (ed.): The Mockers and the Mocked: Comparative Perspectives on Differentiation, Convergence and Diversity in Higher Education. Pergamon Press, Oxford.
Wielemans, W./Vanderhoeven, J.L. (1991): Market Impact and Policy Drift: Belgian Higher Education. In: Neave, G./Van Vught, F. (eds.): Prometheus Bound: The Changing Relationship Between Government and Higher Education in Western Europe. Pergamon Press, Oxford.

Music and Evolution
DNA and the Evolution of Motifs in Beethovens greatest Piano Work
"The Hammerklavier Sonata"

Michael Leslie

The DNA

The "Hammerklavier Sonata" is Beethoven's greatest, longest and most demanding piano sonata. In actual size it is exceeded only by the "33 Variations on a Waltz by Anton Diabelli", his last great work for piano. For his contemporaries the "Hammerklavier Sonata" was monstrously long and hideously difficult with an unplayable, incomprehensible finale.

Because of its enormous length Beethoven took extreme care over the organisation of his material. Many of his largest, most complex works evolved from tiny melodic cells. This phenomenon can be seen for example in the first movements of the 5th and 9th symphonies. During the two years prior to the composition of the "Hammerklavier Sonata" he explored the device of deriving all the subsequent main themes of a work from the very opening bars. In the piano sonata opus 101 and the sonata for violoncello and piano opus 102/1 he employed this strategy and underlined the procedure by repeating the opening bars as a kind of flashback just before the start of the finale.

For the "Hammerklavier Sonata" he hit on an idea which is both simpler and more complex. Throughout the entire length of the work he employed thirds in a way which no other composer either before or after ever attempted, nor, for that matter, did he ever again use this method himself. All the major themes in all the movements are built on a framework of thirds. Thirds appear also in key relationships between large sections within a movement. In addition, all movements of the work (including the introduction to the fugue) abound in passages which are built on seemingly endless chains of descending thirds.

As an example of a theme built on thirds let us take the opening theme of the first movement:

Thirds in the form of (sub)mediant relationships determine the course of all four movements. Again taking the first movement as our example: After the principal theme is in B Flat Major, the entire second group appears in G Major (a third below); the main fugal section of the development section is in E Flat Major (again a third below) and the conclusion of the development section is in B Major (down a third again) causing a semitonal clash with the beginning of the recapitulation in B Flat Major. This tonic – flattened supertonic clash is an integral part of the design of the first, second and third movements.

In the fugue it is not quite so evident, but the point is nonetheless made by the appearance in the key of B Minor of a ghostly episode involving a retrograde version of the fugue subject. One has the impression that with the reversal of the theme a negation of it is intended, a view supported by a scrawled remark in one of Beethoven's sketch books: "B Minor – black key"

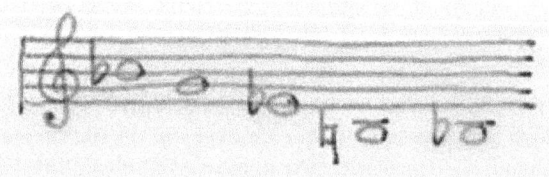

Finally, as an example of a passage in which chains of descending thirds act as a driving force a theme taken from the second group in the first movement:

Music and Evolution 173

For the sake of convenience the examples quoted above have all been taken from the first movement. This was not in fact necessary, since instances abound in all the other movements. Beethoven was a supreme architect and the "Hammerklavier Sonata" is one of his most highly organised works. For a fuller account I refer the reader to "The Classical Style. Haydn, Mozart, Beethoven" by Charles Rosen (The Viking Press, Inc., New York 1971, 1976).

The Evolution of Motifs

Since the publication of "The Classical Style" the material presented above is well known among musicians. The following, however, is an essay of my own, and the discovery it contains is known only to a small circle of friends and colleagues. I present it in its original form.

In the present context, however, I wish to add that the chain of descending thirds referred to in the second paragraph can be followed in the appendix which presents the entire introduction to the fugue. The sequence of thirds is to be found in the bass line. With the exception of the A natural in Episode 8 all the notes constituting the chain of thirds are presented as octaves. In Episode 9 the repeated A natural octaves directly under the word "Prestissimo" contain two other notes, a C sharp and an E natural. In spite of the interruptions along the way it is easy to follow the chain of thirds since the notes constituting the chain are the only bass octaves in the entire introduction. The introduction begins with a series of Fs encompassing almost the full extent of Beethoven's keyboard. In Episode 1 a descent is made from this F to D flat, B flat and G flat. Although it is not restruck, this G flat constitutes the harmonic basis for the whole of Episode 2. In Episode 3 the downward motion is continued: G flat, E flat, C flat (or B natural). Similarly, this B natural provides the harmonic basis for Episode 4. It is sounded again in Episode 5 and moves in Episode 6 to G sharp which again functions as the harmonic basis for the entire episode. Episode 7 continues the descent with G sharp, E natural and C sharp. In Episode 8 A natural is reached. In Episode 9 the motion is greatly accelerated. The steps are:

A natural, F sharp, D, B, G, E, C, A, F, D. The next leap is not a third but a fourth: D to A. From A, however, the bass drops again a third to F, and from here the fugal finale begins.

A New Look at the Introduction to the Fugue of the Beethoven "Hammerklavier" Sonata

As a major landmark in the classical piano literature the "Hammerklavier" Sonata has been subjected to repeated musical analysis and spiritual interpretation [1]. It is widely believed that the "Hammerklavier" Sonata, (written in 1817/18), precipitated the artistic and emotional breakthrough that released Beethoven from the deadlock of the "fallow" years (1812-1817) and enabled him to enter the third, final creative period. The slow movement and fugue in particular exhibit the intricate voice leading, subtle harmonic relationships and polyphonic harshness associated with the last period. Of crucial importance to the work is the tentative, groping transition from the slow movement to the fugue, a passage which Sulli-

---

1   J. W. N. Sullivan: Beethoven, His Spiritual Development; D. F. Tovey: A Companion to Beethoven's Pianoforte Sonatas; J. Uhde: Beethovens Klaviermusik; C. Rosen: The Classical Style: Haydn, Mozart, Beethoven; F. Busoni: Nachtrag zum ersten Teil der Busoni-Ausgabe des Wohltemperierten Klaviers; etc.

Music and Evolution 175

van saw as "a miracle of art" and a record of the soul's journey from fathomless despair to a blind, furious will to live.

Existing analyses of this introduction to the fugue make much of a chain of descending thirds and a general transition from homophony to polyphony. Beyond this, however, they reveal neither a plan nor a coherent purpose. The present examination was undertaken in the belief that a work like the "Hammerklavier" Sonata is organised down to the last semiquaver, and its aim was to discover, if possible, a hidden order in the introduction behind the seemingly random succession of episodes. This order is elucidated in the following paragraphs.

Stated simply, the introduction prefigures, in a fantastical and yet utterly logical way, the theme of the final movement, the subject of the fugue. It would appear that as a first step Beethoven mentally divided the fugue theme into four segments. These segments are here referred to as *Sections A, B, C and D*. The introduction itself consists of nine vignettes (referred to as *Episodes 1-9*). Each of these vignettes (with one exception) is either an encoded version of, or a miniature contrapuntal fantasia based on, one of the *Sections A to D* of the fugue theme. Once the complete theme has been presented in this disguised and piecemeal fashion the introduction comes to an end and the finale can begin. The plan is both simple and ingenious, and probably evolved further to become the design for the beginning of the last movement of the ninth symphony. The reader is referred to the summary, the table on page 8 and the appendix.

The following example shows the four segments *(Sections A, B, C and D)* of the fugue subject.

The first bar of the introduction presents an encoded version of the melodic and harmonic outlines of *Sections A, B, and C* of the fugue theme.

Episodes 1 and 2

Music and Evolution 177

Despite the seamless continuity of the above example I have divided it, for reasons which will become clear, into two episodes: *Episode 1* (from the beginning to the G flat major chord under the pause – *Section A* of the theme) and *Episode 2* (the rest – *Sections B and C).*

Seven episodes follow this "overview" of the theme. Each episode refers to one and only one section of the fugue theme. The first of these episodes, *(Episode 3: from the beginning of bar 2 until the key change)* consists of three tentative, descending broken chords. A glance at the three analagous chords at the beginning of the previous example *(Episode 1)* suffices to show that again here these chords stand for *Section A* of the theme (the upward leap of a tenth and the trill).

Episode 3:

Like the Promenade in Moussorgsky's "Pictures at an Exhibition" this episode returns in slightly varied form as a connecting link between the larger episodes. It invariably refers back to Section A and shall henceforth be ignored.

The next episode – in B major – is derived from *Section B:*

Episode 4, in B major

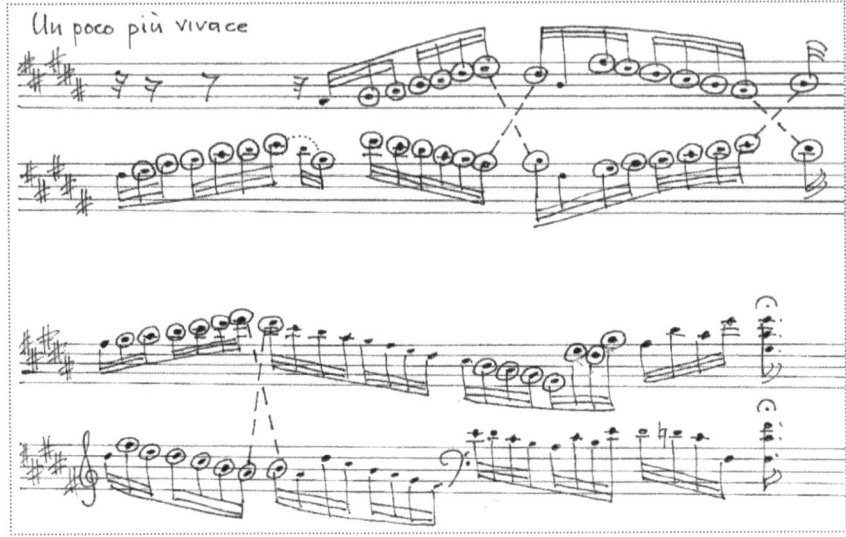

Episode 6, in G sharp minor, is built around *Section D* of the theme. Beethoven focuses here on two motives – a fast "fluttering" figure and a more slowly moving section of descending scale:

Section D

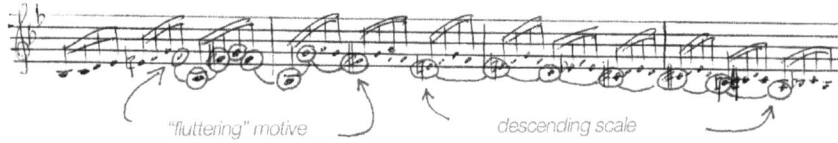

Music and Evolution

This can most clearly be seen in the lower voice in the first two bars:

It is not quite so easy to discern a correspondence between *Section D* of the theme and *Episode 8*.

Episode 8

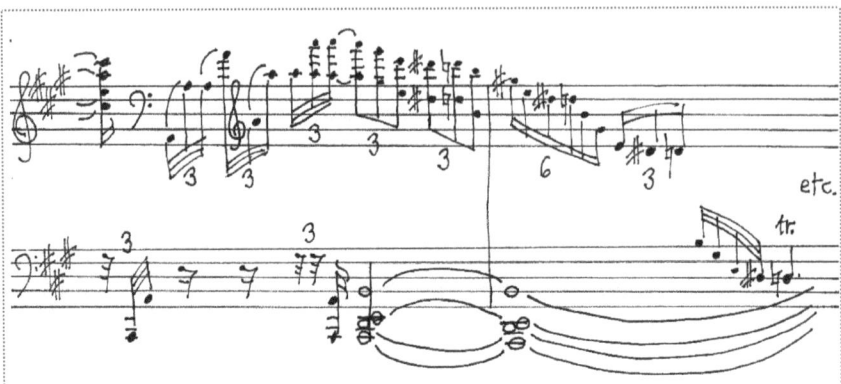

Section C

An (admittedly rather weak) common aspect can be seen in the following succession of chromatically descending tritone intervals in *Episode 8:*

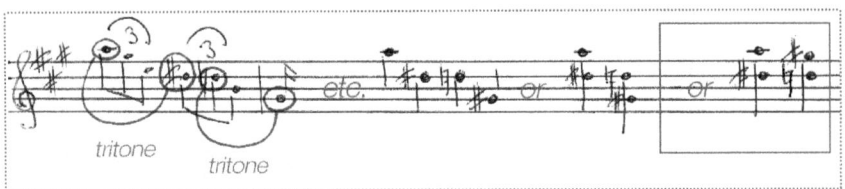

and in *Section C* of the theme:

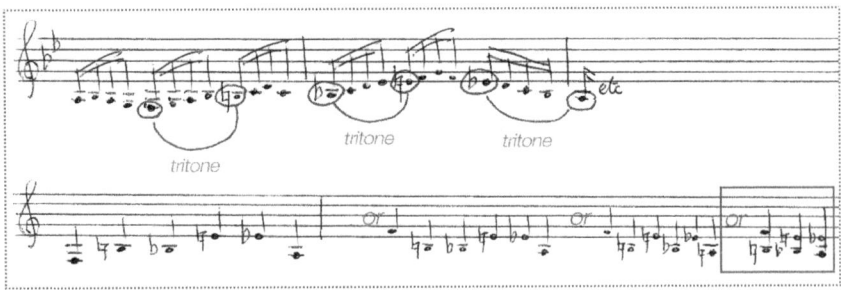

In fact the whole of *Episode C* is laced with these tritone leaps:

Nonetheless this link is weak. In the theme the tritone sequence is structural, energetic and clearly audible, whereas it appears almost accidental by comparison in the episode. However, new, promising vistas open when the episode is transposed up a semitone:

This clearly recalls bars 307 and 308 of the first movement:

and at the same time the connection to the fugue subject becomes obvious:

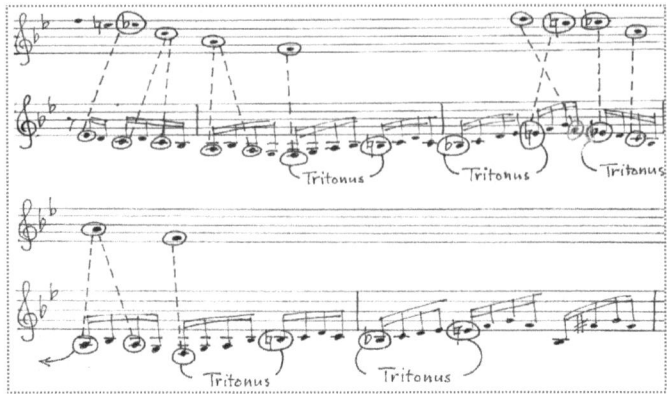

Despite its weaknesses the tritone connection has been preserved in the above example. It fills in the gaps which occur when the fugue subject rises.
The following passage,

Episode 9,

is a new version of *Episode 1*. Since all four sections of the fugue theme have now been presented, there is nothing further for this recurring episode to connect up with: it has reached the end of its road. What now happens is very interesting. The music behaves exactly like a clockwork mechanism which is been removed from its encasement, (and similarly has nothing left to connect up to). It gathers speed, rears up, hammers away frantically and finally sinks down exhausted. At this moment the fugue theme finally appears in its true form like a phoenix rising from the ashes and the finale can begin.

An analogy will best describe the psychological thrust of the introduction to the fugue. The dreamlike episodes of this passage appear as a kind of musical equivalent of those bizarre visions that arise in the borderline state between sleeping and waking and drift in seemingly random succession before the sleeper's inward eye. Here, however, in a crescendo of mounting tension their latent energy and concealed meaning force an entry into the sleeper's consciousness, (in the "clockwork" passage just described). At the instant of waking the dreamer spontaneously apprehends the inner significance of the dream sequence. The underlying, hitherto slumbering idea (the fugue subject) leaps naked and resolute into action. (Students of organic chemistry will recognize here a creative process identical to that so vividly described by Friedrich Kekulé in connection with the moment of his discovery of the molecular structure of the benzene ring).

To recapitulate: Beethoven divided the fugue subject into four segments: *A, B, C,* and *D:*

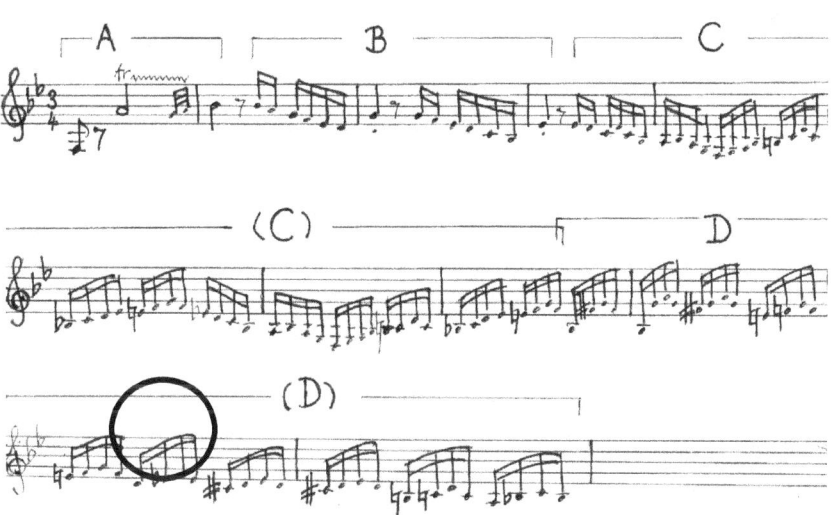

Recondite and encoded as in a dream, all these sections of the theme are presented, one after another, in a series of episodes until the complete thematic material lies before our eyes. Following an "overview" of the greater part of the theme, *(Episode 1, Section A; – Episode 2, Sections B and C encircled in red below),* all the various sections of the theme are presented singly, one section to one episode, in logical order, with *Episode 1 (Section A)* functioning as a connecting link. (For dramatic reasons the order of appearance of *Sections C and D* is reversed).

A   B C   A   B   A   D   A   C   A

Bar (Introduction)	Episode	Section of the Fuge Theme
1	1	Section A, (Bar 1)
1	2	Section B & C, (Bar 2 – 8)
2	3	Section A, (Bar 1)
2	4	Section B, (Bar 2 – 3)
2	5	Section A, (Bar 1)
3-7	6	Section D, (Bar 8 – 11)
8	7	Section A, (Bar 1)
9-10	8	Section C, (Bar 4 – 8)
10	9	Section A, (Bar 1)

One of the most intriguing aspects of this introduction is surely the fact that Beethoven was so obviously writing here for the initiate rather than for the unprepared listener, no matter how perceptive he or she may be. Without prior knowledge a listener cannot possibly understand what is going on here; the relationship of the episodes to the sections of the fugue theme simply cannot be heard.

(Obviously, however, the introduction can still be experienced as stunningly wayward and psychologically satisfying music. This is of course the way it usually is heard). On the other hand, an appreciation of the marvellous structures and correspondences here outlined changes the player's perception (and thereby automatically his performance), and a listener who already knows what to look out for will enjoy both the enhanced interpretation and Beethoven's elegant psychological game. But there is no escaping the fact that he must already know what he is about to hear, in order to be able to hear it.

Music and Evolution 185

Appendix: Introduction to the Fugue of the "Hammerklavier" Sonata

Representation of the episodes of the introduction and the segments of the fugue subject

Urtext Henle-Edition 1953

Biographies

Francisco J. Ayala is University Professor and Donald Bren Professor of Biological Sciences, and Professor of Philosophy at the University of California, Irvine.

Sarah Chan is Deputy Director of the ISEI (Institute for Science, Ethics and Innovation) and Research Fellow in Bioethics and Law at the University of Manchester.

Mikhail Naumovich Epstein is Samuel Chandler Dobbs Professor of Cultural Theory and Russian Literature at Emery University (USA) and Professor of Russian and Cultural Theory and Director of Centre for Humanities Innovation at Durham University (UK).

Nikola Grimm studies medicine and philosophy and is currently making research in the field of moral enhancement at the Institute of History and Ethics of Medicine at the Friedrich-Alexander-University Erlangen/Nuernberg.

Hille Haker is the Richard McCormick, S. J., Chair of Moral Theology at Loyola University Chicago.

Ottfried Höffe is Professor em. of Philosophy at the University in Tuebingen.

Branka-Rista Jovanovic has studied philosophy at the University of Belgrade and Munich and has been working for several NGOs.

Nikolaus Knoepffler is full Professor of Applied Ethics at the Friedrich-Schiller-University Jena and Founder and President of the Global Applied Ethics Network.

Michael Leslie is an Australian concert pianist living in Munich.

Dietmar Mieth is Professor em. of Theological Ethics / Social Ethics at the Catholic Theological Faculty of the University of Tuebingen, and since 2009 Fellow at the Max Weber Institute for Advanced Studies at the University of Erfurt.

Vojin Rakić is Director of the Center for the Study of Bioethics, Institute for Philosophy and Social Theory, University of Belgrade.

Stefan Lorenz Sorgner has a Dr. phil. in philosophy and is Lecturer of Medical Ethics at the Institute of History and Ethics of Medicine at the Friedrich-Alexander-University Erlangen/Nuernberg.

BEYOND HUMANISM: TRANS- AND POSTHUMANISM
JENSEITS DES HUMANISMUS: TRANS- UND POSTHUMANISMUS

Edited by / Herausgegeben von Stefan Lorenz Sorgner

Vol./Bd. 1 Robert Ranisch / Stefan Lorenz Sorgner (eds.): Post- and Transhumanism. An Introduction. *In preparation.*

Vol./Bd. 2 Stephen R. L. Clark: Philosophical Futures. 2011.

Vol./Bd. 3 Hava Tirosh-Samuelson / Kenneth L. Mossman (eds.): Building Better Humans? Refocusing the Debate on Transhumanism. 2012.

Vol./Bd. 4 Elizabeth Butterfield: Sartre and Posthumanist Humanism. 2012.

Vol./Bd. 5 Stefan Lorenz Sorgner / Branka-Rista Jovanovic (eds.). In cooperation with Nikola Grimm: Evolution and the Future. Anthropology, Ethics, Religion. 2013.

www.peterlang.de

www.ingramcontent.com/pod-product-compliance
Ingram Content Group UK Ltd.
Pitfield, Milton Keynes, MK11 3LW, UK
UKHW041913140426
5217IPUK00002B/19